Use R!

Series Editors:
Robert Gentleman   Kurt Hornik   Gi

For further volumes:
http://www.springer.com/series/6991

Dirk Eddelbuettel

Seamless R and C++ Integration with Rcpp

 Springer

Dirk Eddelbuettel
River Forest
Illinois, USA

ISBN 978-1-4614-6867-7   ISBN 978-1-4614-6868-4 (eBook)
DOI 10.1007/978-1-4614-6868-4
Springer New York Heidelberg Dordrecht London

Library of Congress Control Number: 2013933242

© The Author 2013
This work is subject to copyright. All rights are reserved by the Publisher, whether the whole or part of the material is concerned, specifically the rights of translation, reprinting, reuse of illustrations, recitation, broadcasting, reproduction on microfilms or in any other physical way, and transmission or information storage and retrieval, electronic adaptation, computer software, or by similar or dissimilar methodology now known or hereafter developed. Exempted from this legal reservation are brief excerpts in connection with reviews or scholarly analysis or material supplied specifically for the purpose of being entered and executed on a computer system, for exclusive use by the purchaser of the work. Duplication of this publication or parts thereof is permitted only under the provisions of the Copyright Law of the Publisher's location, in its current version, and permission for use must always be obtained from Springer. Permissions for use may be obtained through RightsLink at the Copyright Clearance Center. Violations are liable to prosecution under the respective Copyright Law.
The use of general descriptive names, registered names, trademarks, service marks, etc. in this publication does not imply, even in the absence of a specific statement, that such names are exempt from the relevant protective laws and regulations and therefore free for general use.
While the advice and information in this book are believed to be true and accurate at the date of publication, neither the authors nor the editors nor the publisher can accept any legal responsibility for any errors or omissions that may be made. The publisher makes no warranty, express or implied, with respect to the material contained herein.

Springer is part of Springer Science+Business Media (www.springer.com)

To Lisa, Anna and Julia

Preface

Rcpp is an R add-on package which facilitates extending R with C++ functions.

It is being used for anything from small and quickly constructed add-on functions written either to fluidly experiment with something new or to accelerate computing by replacing an R function with its C++ equivalent to large-scale bindings for existing libraries, or as a building block in entirely new research computing environments.

While still relatively new as a project, **Rcpp** has already become widely deployed among users and developers in the R community. **Rcpp** is now the most popular language extension for the R system and used by over 100 CRAN packages as well as ten BioConductor packages.

This books aims to provide a solid introduction to **Rcpp**.

Target Audience

This book is for R users who would like to extend R with C++ code. Some familiarity with R is certainly helpful; a number of other books can provide refreshers or specific introductions. C++ knowledge is also helpful, though not strictly required. An appendix provides a very brief introduction for C++ to those familiar only with the R language.

The book should also be helpful to those coming to R with more of a C++ programming background. However, additional background reading may be required to obtain a firmer grounding in R itself. Chambers (2008) is a good introduction to the philosophy behind the R system and a helpful source in order to acquire a deeper understanding.

There may also be some readers who would like to see how **Rcpp** works internally. Covering that aspect, however, requires a fairly substantial C++ content and is not what this book is trying to provide. The focus of this book is clearly on how to *use* **Rcpp**.

Historical Context

Rcpp first appeared in 2005 as a (fairly small when compared to its current size) contribution by Dominick Samperi to the **RQuantLib** package started by Eddelbuettel in 2002 (Eddelbuettel and Nguyen 2012). **Rcpp** became a CRAN package in its own name in early 2006. Several releases (all provided by Samperi) followed in quick succession under the name **Rcpp**. The package was then renamed to **RcppTemplate**; several more releases followed during 2006 under the new name. However, no new releases were made during 2007, 2008, or most of 2009. Following a few updates in late 2009, the **RcppTemplate** package has since been archived on CRAN for lack of active maintenance.

Given the continued use of the package, Eddelbuettel decided to revitalize it. New releases, using the original name **Rcpp**, started in November 2008. These included an improved build and distribution process, additional documentation, and new functionality—while retaining the existing "classic **Rcpp**" interface. While not described here, this API will continue to be provided and supported via the **Rcpp-Classic** package (Eddelbuettel and François 2012c).

Reflecting evolving C++ coding standards (see Meyers 2005), Eddelbuettel and François started a significant redesign of the code base in 2009. This added numerous new features, many of which are described in the package via different vignettes. This redesigned version of **Rcpp** (Eddelbuettel and François 2012a) has become widely used with over ninety CRAN packages depending on it as of November 2012. It is also the version described in this book.

Rcpp continues to be under active development, and extensions are being added. The content described here shall remain valid and supported.

Related Work

Integration of C++ and R has been addressed by several authors; the earliest published reference is probably Bates and DebRoy (2001). The "Writing R Extensions" manual (R Development Core Team 2012d) has also been mentioning C++ and R integration since around that time. An unpublished paper by Java et al. (2007) expresses several ideas that are close to some of our approaches, though not yet fully fleshed out. The **Rserve** package (Urbanek 2003, 2012) acts as a socket server for R. On the server side, **Rserve** translates R data structures into a binary serialization format and uses TCP/IP for transfer. On the client side, objects are reconstructed as instances of Java or C++ classes that emulate the structure of R objects.

The packages **rcppbind** (Liang 2008), **RAbstraction** (Armstrong 2009a), and **RObjects** (Armstrong 2009b) are all implemented using C++ templates. None of them have matured to the point of a CRAN release. **CXXR** (Runnalls 2009) approaches this topic from the other direction: its aim is to completely refactor R on a stronger C++ foundation. **CXXR** is therefore concerned with all aspects of the R interpreter, read-eval-print loop (REPL), and threading; object interchange between R and C++ is but one part. A similar approach is discussed by Temple Lang (2009a) who suggests making low-level internals extensible by package developers in order to facilitate extending R. Temple Lang (2009b), using compiler output for references on the code in order to add bindings and wrappers, offers a slightly different angle. Lastly, the **rdyncall** package (Adler 2012) provides a direct interface from R into C language APIs. This can be of interest if R programmers want to access lower-level programming interfaces directly. However, it does not aim for the same object-level interchange that is possible via C++ interfaces, and which we focus on with **Rcpp**.

Typographic Convention

The typesetting follows the usage exemplified both by the publisher, and by the *Journal of Statistical Software*. We use

- Sans-serif for programming language such as R or C++
- Boldface for (CRAN or other) software packages such as **Rcpp** or **inline**
- Courier for short segments of code or variables such as `x <- y + z`

We make use of a specific environment for the short pieces of source code interwoven with the main text.

River Forest, IL, USA   Dirk Eddelbuettel

Acknowledgements

Rcpp is the work of many contributors, and a few words of thanks are in order.

Dominick Samperi contributed the original code which, while much more limited in scope than the current **Rcpp**, pointed clearly in the right direction of using C++ templates to convert between R and C++ types.

Romain François has shown impeccable taste in designing and implementing very large parts of **Rcpp** as it is today. The power of the current design owes a lot to his work and boundless energy. Key components such as modules and sugar, as well as lot a of template "magic," are his contributions. This started as an aside to make object interchange easier for our **RProtoBuf** package—and it has taken us down a completely different, but very exciting road. It has been a pleasure to work with Romain, I remain in awe of his work, and I look forward to many more advances with **Rcpp**.

Doug Bates has been a help from the very beginning: had it not been for some simple macros to pick list components out of SEXP types, I may never have started **RQuantLib** a decade ago. Doug later joined this project and has been instrumental in a few key decisions regarding **Rcpp** and **RcppArmadillo** and has taken charge of the **RcppEigen** project.

John Chambers become a key supporter right when *Rcpp modules* started and contributed several important pieces at the gory intersection between R and C++. It was very flattering for Romain and me to hear from John how **Rcpp** is so close to an original design vision of a whole-object interchange between systems which was already present on a hand-drawn Bell Labs designs from the 1970s.

JJ Allaire has become a very important contributor to **Rcpp** and a key supporter of the same idea of an almost natural pairing between R and C++. The *Rcpp attributes* which he contributed are showing a lot of promise, and we expect great things to be built on top of this.

Several other members of the R Core team—notably Kurt Hornik, Uwe Ligges, Martyn Plummer, Brian Ripley, Luke Tierney, and Simon Urbanek—have helped at various points with anything from build issues and portability to finer points of R internals. Last but not least, there would of course be no **Rcpp** if there was no R system to build upon and to extend.

Finally, many members of the R and **Rcpp** communities have been very supportive at different workshops, conference presentations, and via the mailing lists. Numerous good questions and suggestions have come this way. And, of course, it is seeing this work being used so actively which motivates us and keeps us moving forward with **Rcpp**.

Contents

Part I Introduction

1 A Gentle Introduction to Rcpp 3
  1.1 Background: From R to C++ 3
  1.2 A First Example 7
    1.2.1 Problem Setting 7
    1.2.2 A First R Solution 7
    1.2.3 A First C++ Solution 8
    1.2.4 Using Inline 9
    1.2.5 Using Rcpp Attributes 11
    1.2.6 A Second R Solution 12
    1.2.7 A Second C++ Solution 12
    1.2.8 A Third R Solution 14
    1.2.9 A Third C++ Solution 14
  1.3 A Second Example 15
    1.3.1 Problem Setting 15
    1.3.2 R Solution 15
    1.3.3 C++ Solution 16
    1.3.4 Comparison 17
  1.4 Summary 18

2 Tools and Setup 19
  2.1 Overall Setup 19
  2.2 Compilers 20
    2.2.1 General Setup 20
    2.2.2 Platform-Specific Notes 21
  2.3 The R Application Programming Interface 22
  2.4 A First Compilation with Rcpp 23
  2.5 The Inline Package 25
    2.5.1 Overview 25
    2.5.2 Using Includes 27

		2.5.3 Using Plugins	29
		2.5.4 Creating Plugins	30
	2.6	Rcpp Attributes	31
	2.7	Exception Handling...	32

Part II Core Data Types

3	**Data Structures: Part One**	39
	3.1 The RObject Class..	39
	3.2 The IntegerVector Class	41
	3.2.1 A First Example: Returning Perfect Numbers	42
	3.2.2 A Second Example: Using Inputs	43
	3.2.3 A Third Example: Using Wrong Inputs	44
	3.3 The NumericVector Class.......................................	45
	3.3.1 A First Example: Using Two Inputs	45
	3.3.2 A Second Example: Introducing `clone`	46
	3.3.3 A Third Example: Matrices	47
	3.4 Other Vector Classes	48
	3.4.1 LogicalVector	48
	3.4.2 CharacterVector	49
	3.4.3 RawVector	49

4	**Data Structures: Part Two**	51
	4.1 The Named Class	51
	4.2 The List aka GenericVector Class	52
	4.2.1 List to Retrieve Parameters from R	53
	4.2.2 List to Return Parameters to R	54
	4.3 The DataFrame Class	55
	4.4 The Function Class	56
	4.4.1 A First Example: Using a Supplied Function	56
	4.4.2 A Second Example: Accessing an R Function	56
	4.5 The Environment Class...	57
	4.6 The S4 Class	58
	4.7 ReferenceClasses...	59
	4.8 The R Mathematics Library Functions	60

Part III Advanced Topics

5	**Using Rcpp in Your Package**	65
	5.1 Introduction	65
	5.2 Using `Rcpp.package.skeleton`	66
	5.2.1 Overview	66
	5.2.2 R Code	67
	5.2.3 C++ Code	68
	5.2.4 DESCRIPTION	69
	5.2.5 Makevars and Makevars.win	69

		5.2.6	NAMESPACE	71
	5.3		Case Study: The **wordcloud** Package	73
	5.4		Further Examples	74

6 Extending Rcpp ... 75
- 6.1 Introduction ... 75
- 6.2 Extending Rcpp::wrap ... 76
  - 6.2.1 Intrusive Extension ... 76
  - 6.2.2 Nonintrusive Extension ... 77
  - 6.2.3 Templates and Partial Specialization ... 78
- 6.3 Extending Rcpp::as ... 78
  - 6.3.1 Intrusive Extension ... 78
  - 6.3.2 Nonintrusive Extension ... 79
  - 6.3.3 Templates and Partial Specialization ... 79
- 6.4 Case Study: The **RcppBDT** Package ... 80
- 6.5 Further Examples ... 82

7 Modules ... 83
- 7.1 Motivation ... 83
  - 7.1.1 Exposing Functions Using **Rcpp** ... 83
  - 7.1.2 Exposing Classes Using Rcpp ... 84
- 7.2 Rcpp Modules ... 86
  - 7.2.1 Exposing C++ Functions Using Rcpp Modules ... 86
  - 7.2.2 Exposing C++ Classes Using Rcpp Modules ... 90
- 7.3 Using Modules in Other Packages ... 98
  - 7.3.1 Namespace Import/Export ... 98
  - 7.3.2 Support for Modules in Skeleton Generator ... 99
  - 7.3.3 Module Documentation ... 100
- 7.4 Case Study: The **RcppCNPy** Package ... 100
- 7.5 Further Examples ... 102

8 Sugar ... 103
- 8.1 Motivation ... 103
- 8.2 Operators ... 105
  - 8.2.1 Binary Arithmetic Operators ... 105
  - 8.2.2 Binary Logical Operators ... 106
  - 8.2.3 Unary Operators ... 106
- 8.3 Functions ... 107
  - 8.3.1 Functions Producing a Single Logical Result ... 107
  - 8.3.2 Functions Producing Sugar Expressions ... 107
  - 8.3.3 Mathematical Functions ... 113
  - 8.3.4 The d/q/p/q Statistical Functions ... 114
- 8.4 Performance ... 115
- 8.5 Implementation ... 116

		8.5.1 The Curiously Recurring Template Pattern 117
		8.5.2 The VectorBase Class 117
		8.5.3 Example: sapply 118
	8.6	Case Study: Computing π Using *Rcpp sugar* 122

Part IV Applications

9 RInside 127
    9.1 Motivation 127
    9.2 A First Example: Hello, World! 128
    9.3 A Second Example: Data Transfer 131
    9.4 A Third Example: Evaluating R Expressions 132
    9.5 A Fourth Example: Plotting from C++ via R 133
    9.6 A Fifth Example: Using RInside Inside MPI 134
    9.7 Other Examples 135

10 RcppArmadillo 139
    10.1 Overview 139
    10.2 Motivation: FastLm 140
        10.2.1 Implementation 140
        10.2.2 Performance Comparison 142
        10.2.3 A Caveat 144
    10.3 Case Study: Kalman Filter Using **RcppArmadillo** 146
    10.4 RcppArmadillo and Armadillo Differences 152

11 RcppGSL 155
    11.1 Introduction 155
    11.2 Motivation: FastLm 156
    11.3 Vectors 158
        11.3.1 **GSL** Vectors 158
        11.3.2 RcppGSL::vector 159
        11.3.3 Mapping 161
        11.3.4 Vector Views 161
    11.4 Matrices 163
        11.4.1 Creating Matrices 163
        11.4.2 Implicit Conversion 163
        11.4.3 Indexing 163
        11.4.4 Methods 164
        11.4.5 Matrix Views 164
    11.5 Using **RcppGSL** in Your Package 164
        11.5.1 The `configure` Script 165
        11.5.2 The `src` Directory 166
        11.5.3 The `R` Directory 167
    11.6 Using **RcppGSL** with **inline** 168
    11.7 Case Study: **GSL**-Based B-Spline Fit Using **RcppGSL** 169

12 RcppEigen 177
- 12.1 Introduction 177
- 12.2 Eigen classes 178
  - 12.2.1 Fixed-Size Vectors and Matrices 178
  - 12.2.2 Dynamic-Size Vectors and Matrices 179
  - 12.2.3 Arrays for Per-Component Operations 180
  - 12.2.4 Mapped Vectors and Matrices and Special Matrices 181
- 12.3 Case Study: Kalman filter using RcppEigen 182
- 12.4 Linear Algebra and Matrix Decompositions 183
  - 12.4.1 Basic Solvers 183
  - 12.4.2 Eigenvalues and Eigenvectors 184
  - 12.4.3 Least-Squares Solvers 185
  - 12.4.4 Rank-Revealing Decompositions 185
- 12.5 Case Study: C++ Factory for Linear Models in **RcppEigen** ... 186

Part V Appendix

A **C++ for R Programmers** 195
- A.1 Compiled Not Interpreted 195
- A.2 Statically Typed 197
- A.3 A Better C 198
- A.4 Object-Oriented (But Not Like S3 or S4) 200
- A.5 Generic Programming and the STL 201
- A.6 Template Programming 203
- A.7 Further Reading on C++ 204

References 207

Subject Index 211

Software Index 217

Author Index 219

List of Tables

Table 1.1	Run-time performance of the recursive Fibonacci examples	10
Table 1.2	Run-time performance of the different VAR simulation implementations	17
Table 8.1	Run-time performance of *Rcpp sugar* compared to R and manually optimized C++	116
Table 8.2	Run-time performance of *Rcpp sugar* compared to R for simulating π	124
Table 11.1	Correspondence between **GSL** vector types and templates defined in **RcppGSL**	161
Table 11.2	Correspondence between **GSL** vector view types and templates defined in **RcppGSL**	162
Table 12.1	Mapping between **Eigen** matrix and vector types, and corresponding array types	181
Table 12.2	`lmBenchmark` results for the **RcppEigen** example	191

List of Figures

Figure 1.1   Plotting a density in R 5
Figure 1.2   Plotting a density and bootstrapped confidence interval in R . 6
Figure 1.3   Fibonacci spiral based on first 34 Fibonacci numbers 8

Figure 9.1   Combining **RInside** with the Qt toolkit
for a GUI application 135
Figure 9.2   Combining **RInside** with the Wt toolkit
for a web application 136

Figure 10.1   Object trajectory and Kalman filter estimate 149

Figure 11.1   Artificial data and B-spline fit 175

List of Listings

1.1	Plotting a density in R	4
1.2	Plotting a density and bootstrapped confidence interval in R	4
1.3	Fibonacci number in R via recursion	7
1.4	Fibonacci number in C++ via recursion	8
1.5	Fibonacci wrapper in C++	9
1.6	Fibonacci number in C++ via recursion, using inline	9
1.7	Fibonacci number in C++ via recursion, using Rcpp attributes	11
1.8	Fibonacci number in C++ via recursion, via Rcpp attributes and `sourceCpp`	11
1.9	Fibonacci number in R via memoization	12
1.10	Fibonacci number in C++ via memoization	12
1.11	Fibonacci number in R via iteration	14
1.12	Fibonacci number in C++ via iteration	14
1.13	VAR(1) of order 2 generation in R	16
1.14	VAR(1) of order 2 generation in C++	16
1.15	Comparison of VAR(1) run-time between R and C++	17
2.1	A first manual compilation with Rcpp	23
2.2	A first manual compilation with Rcpp using Rscript	24
2.3	Using the first manual compilation from R	24
2.4	Convolution example using inline	26
2.6	Using inline with `include=`	27
2.5	Program source from convolution example using inline in verbose mode	28
2.7	A first RcppArmadillo example for inline	29
2.8	Creating a plugin for use with inline	30
2.9	Example of new `cppFunction`	31
2.10	Example of new `cppFunction` with plugin	32
2.11	C++ example of throwing and catching an exception	32
2.12	Using C++ example of throwing and catching an exception	33
2.13	C++ example of example from Rcpp-type checks	33
2.14	C++ macros for Rcpp exception handling	34

2.15	inline version of C++ example of throwing and catching an exception	34
2.16	Rcpp attributes version of C++ example of throwing and catching an exception	34
3.1	A function to return four perfect numbers	42
3.2	A function to reimplement `prod()`	43
3.3	A second function to reimplement `prod()`	43
3.4	Testing the `prod()` function with floating-point inputs	44
3.5	Testing the `prod()` function with inappropriate inputs	45
3.6	A function to return a generalized sum of powers	45
3.7	Declaring two vectors from the same `SEXP` type	46
3.8	Declaring two vectors from the same `SEXP` type using clone	46
3.9	Using Rcpp sugar to compute a second vector	47
3.10	Declaring a three-dimensional vector	47
3.11	A function to take square roots of matrix elements	48
3.12	A function to assign a logical vector	48
3.13	A function to assign a character vector	49
4.1	A named vector in R	51
4.2	A named vector in C++	52
4.3	A named vector in C++, second approach	52
4.4	Using the `List` class for parameters	53
4.5	Using a `List` to return objects to R	54
4.6	Using the `DataFrame` class	55
4.7	Using a `Function` passed as argument	56
4.8	Using a `Function` accessed from R	57
4.9	Using a `Function` via an `Environment`	57
4.10	Assigning in the global environment	58
4.11	A simple example for accessing S4 class elements	58
4.12	A simple example for accessing S4 class elements	59
4.13	Example use of `Rmath.h` functions	60
5.1	A first `Rcpp.package.skeleton` example	66
5.2	Files created by `Rcpp.package.skeleton`	67
5.3	R function `rcpp_hello_world`	67
5.4	C++ header file `rcpp_hello_world.h`	68
5.5	C++ source file `rcpp_hello_world.cpp`	68
5.6	Calling R function `rcpp_hello_world`	69
5.7	DESCRIPTION file for skeleton package	69
5.8	Makevars file for skeleton package	70
5.9	Makevars.win file for skeleton package	70
5.10	NAMESPACE file for skeleton package	71
5.11	Manual page `mypackage-package.Rd` for skeleton package	71
5.12	Manual page `rcpp_hello_world.Rd` for skeleton package	72
5.13	Function `is_overlap` from the wordcloud package	73
6.1	as and wrap declarations	75
6.2	Implicit use of as and wrap	75

List of Listings

6.3	Intrusive extension for `wrap`	77
6.4	Nonintrusive extension for `wrap`	77
6.5	Partial specialization for `wrap`	78
6.6	Intrusive extension for `as`	78
6.7	Nonintrusive extension for `as`	79
6.8	Partial specialization via `Exporter`	79
6.9	Partial specialization of `as` via `Exporter`	80
6.10	RcppBDT definitions of `as` and `wrap`	81
6.11	RcppBDT use of `as` and `wrap`	81
6.12	RcppBDT example for `getFirstDayOfWeekAfter`	82
7.1	A simple `norm` function in C++	84
7.2	Calling the `norm` function	84
7.3	A simple class `Uniform`	84
7.4	Exposing two member functions for `Uniform` class	85
7.5	Using the `Uniform` class from R	86
7.6	Exposing the `norm` function via modules	86
7.7	Using `norm` function exposed via modules	87
7.8	A module example with six functions	87
7.9	Modules example interface	87
7.10	Modules example use from R	88
7.11	Modules example with function documentation	88
7.12	Output for modules example with function documentation	89
7.13	Modules example with documentation and formal arguments	89
7.14	Output for modules example with documentation and formal arguments	89
7.15	Modules example with documentation and formal arguments without defaults	89
7.16	Usage of modules example with documentation and formal arguments	90
7.17	Modules example with ellipis argument	90
7.18	Output of modules example with ellipis argument	90
7.19	Exposing `Uniform` class using modules	91
7.20	Using `Uniform` class via modules	91
7.21	Constructor with a description	92
7.22	Constructor with a validator function pointer	92
7.23	Exposing fields and properties for modules	92
7.24	Field with documentation	93
7.25	Readonly-field with documentation	93
7.26	Property with getter and setter, or getter-only	93
7.27	Example of using a getter	94
7.28	Example of using a setter	94
7.29	Example code for properties	94
7.30	Example using properties	95
7.31	Example documenting a method	95
7.32	Const and non-const member functions	96

7.33	Example of S4 dispatch	96
7.34	Complete example of exposing `std::vector<double>`	97
7.35	R use of `std::vector<double>` modules example	98
7.36	R NAMESPACE import of Rcpp for modules	98
7.37	R `.onLoad()` code for module	98
7.38	Package skeleton support for modules	99
7.39	Use of `prompt` for documentation skeleton	100
7.40	`NumPy` load and save functions defined in RcppCNPy	100
7.41	Example of module declaration in RcppCNPy	102
8.1	A simple C++ function operating on vectors	103
8.2	A simple R function operating on vectors	104
8.3	A simple C++ function using sugar operating on vectors	104
8.4	Binary arithmetic operators for sugar	105
8.5	Binary logical operators for sugar	106
8.6	Unary operators for sugar	106
8.7	Functions returning a single boolean result	107
8.8	Using functions returning a single boolean result	107
8.9	Example using `is_na` sugar function	108
8.10	Example using `seq_along` sugar function	108
8.11	Example using `seq_len` sugar function	108
8.12	Example using `pmin` and `pmax` sugar function	109
8.13	Example using `ifelse` sugar function	109
8.14	Example using `sapply` sugar function	109
8.15	Example using `std::unary_function` functor with `sapply`	110
8.16	Example using `std::unary_function` functor with `mapply`	110
8.17	Example using `sign` sugar function	111
8.18	Example using `sign` sugar function	111
8.19	Example using `setdiff` sugar function	111
8.20	Example using `union_` sugar function	111
8.21	Example using `intersect` sugar function	112
8.22	Example using `clamp` sugar function	112
8.23	Example using `unique` sugar function	112
8.24	Example using `table` sugar function	113
8.25	Example using `duplicated` sugar function	113
8.26	Examples using mathematical sugar functions	113
8.27	Examples of d/p/q/r statistical sugar functions sugar	114
8.28	Examples of using sugar RNG functions with `RNGScope`	115
8.29	The Curiously Recurring Template Pattern (CRTP)	117
8.30	The `VectorBase` class for *Rcpp sugar*	117
8.31	The `sapply` *Rcpp sugar* implementation	119
8.32	`Rcpp::traits::result_of` template	120
8.33	`result_of` trait implementation	120
8.34	`Rcpp::traits::r_sexptype_traits` template	120
8.35	`r_vector_element_converter` class	121
8.36	`storage_type` trait	121

List of Listings

8.37	Input expression base type	121
8.38	Output expression base type	121
8.39	Constructor for `Sapply` class template	122
8.40	Implementation of `Sapply`	122
8.41	Simulating π in R	123
8.42	Simulating π in C++	123
8.43	Simulating π in R	123
9.1	First RInside example: Hello, World!	128
9.2	Makefile for RInside examples	129
9.3	Using Makefile for RInside to build example	130
9.4	Second RInside example: data transfer	131
9.5	Third RInside example: data transfer	132
9.6	Fourth RInside example: plotting from C++ via R	133
9.7	Fifth RInside example: parallel computing with MPI	134
10.1	A simple Armadillo example	139
10.2	`FastLm` function using RcppArmadillo	141
10.3	Basic `fLm()` function without formula interface	142
10.4	Basic `fastLmPure()` R function without formula interface	143
10.5	`FastLm` comparison	143
10.6	An example of a rank-deficient design matrix	144
10.7	Basic Kalman filter in **Matlab**	146
10.8	Basic Kalman filter in R	147
10.9	Basic Kalman filter in R	148
10.10	Basic Kalman filter class in C++ using Armadillo	150
10.11	Basic Kalman filter function in C++	151
10.12	Basic Kalman filter timing comparison	151
10.13	Standard defines for RcppArmadillo	152
10.14	Standard defines for RcppArmadillo	153
11.1	`FastLm` function using RcppGSL	157
11.2	Definition of `gsl_vector` and `gsl_vector_int`	158
11.3	Example use of `gsl_vector`	159
11.4	Example use of `RcppGSL::vector<T>`	159
11.5	Example `RcppGSL::vector<T>` function	160
11.6	Example call of `RcppGSL::vector<T>` function	160
11.7	Second example `RcppGSL::vector<T>` function	160
11.8	Example call of second `RcppGSL::vector<T>` function	161
11.9	Example of a vector view class	162
11.10	Example use RcppGSL matrix class	163
11.11	Implicit conversion for RcppGSL matrix class	163
11.12	Indexing for RcppGSL matrix class	164
11.13	Matrix norm in R	165
11.14	`Autoconf` script for RcppGSL use	165
11.15	Shell script configuration script for RcppGSL use	166
11.16	Windows shell script configuration script for RcppGSL use	166
11.17	Vector norm function for RcppGSL	166

11.18	`Makevars.in` for RcppGSL example	167
11.19	R function for RcppGSL example	168
11.20	Using RcppGSL with inline	168
11.21	Using `package.skeleton` with inline result	169
11.22	B-spline fit example from the GSL	169
11.23	Beginning of C++ file with B-spline fit for R	172
11.24	Data generation for GSL B-spline fit for R	172
11.25	Data fit for GSL B-spline with R	173
11.26	R side of GSL B-spline example	175
12.1	A simple Eigen example using fixed-size vectors and matrices	178
12.2	Eigen fixed-size vector and matrix representation	178
12.3	Eigen dynamic-size vector and matrix representation	179
12.4	A simple Eigen example using dynamic-size vectors and matrices	179
12.5	Comparing performance of simple operations between dynamic and fixed size vectors	179
12.6	Timing results simple operations betweem dynamic and fixed size vectors	180
12.7	Basic Kalman filter class in C++ using Eigen	182
12.8	Using a basic Eigen solver from R	184
12.9	Computing eigenvalues using Eigen	184
12.10	Computing least-squares using Eigen	185
12.11	Rank-revelaing decompositions using Eigen	186
12.12	Core of definition of `lm` class in Eigen	187
12.13	Derived classes of `lm` providing specializations	188
12.14	Implementation of two subclass constructors for `lm` model fit	189
12.15	Selection of subclasses for `lm` model fit	189
12.16	Actual `fastLm` function in `RcppEigen` package	190
A.1	Simple C++ example: Hello, World!	195
A.2	Compiling and linking simple C++ example: Hello, World!	196
A.3	Compiling and linking simple C++ example in one step: Hello, World!	196
A.4	Simple C++ example using Rmath	196
A.5	Compiling and linking simple C++ example using Rmath	197
A.6	Simple R example of dynamic types	197
A.7	Simple R example of dynamic types	198
A.8	Simple C++ function example	199
A.9	Simple C++ function call example	200
A.10	Simple C++ data structure using `struct`	200
A.11	Simple C++ data structure using `class`	201
A.12	Simple C++ example using iterators on `vector`	202
A.13	Simple C++ example using const iterators on `list`	203
A.14	Simple C++ example using const iterators on `deque`	203
A.15	Simple C++ example using `accumulate` algorithm	203
A.16	Simple C++ template example	204
A.17	Another C++ template	204

Part I
Introduction

Chapter 1
A Gentle Introduction to Rcpp

Abstract This initial chapter provides a first introduction to **Rcpp**. It uses a somewhat slower pace and generally more gentle approach than the rest of the book in order to show key concepts which are revisited and discussed in more depth throughout the remainder. So the aim of this chapter is to cover a fairly wide range of material, but at a more introductory level for an initial overview. Two larger examples are studied in detail. We first compute the Fibonacci sequence in three different ways in two languages. Second, we simulate from a multivariate dynamic model provided by a vector autoregression.

1.1 Background: From R to C++

R is both a powerful interactive environment for data analysis, visualization, and modeling and an expressive programming language designed and built to support these tasks. The interactive nature of working with data—through data displays, summaries, model estimation, simulation, and numerous other tasks—is a key strength of the R environment. And, so is the R programming language which permits use from interactive explorations to small scripts and all the way to complete implementations of new functionality. This R programming language is in fact a dialect of the S programming language initially developed by Bell Labs.

The dual nature of interactive analysis, as well as programming, is no accident. As succinctly expressed in the title of one of the books on the S language (which has provided the foundations upon which R is built), it is designed to support *Programming with Data* (Chambers 1998). That is a rather unique proposition as far as programming languages go. As a domain-specific language (DSL), R is tailored specifically to support and enable data analysis work. Moreover, there is also a particular focus on research use for developing new and exciting approaches, as well as solidifying existing approaches. R and its predecessor S are not static languages: they have evolved since the first designs well over thirty years ago and continue to evolve today.

To mention just one example of this evolution, object orientation in R is supported by the S3 and S4 class systems, as well as the newer Reference Classes. Of course, such flexibility of having alternate approaches can also be seen as a weakness. It may lead to yet more material which language beginners may find perplexing, and it may lead to small inconsistencies which may confuse intermediate and advanced users. Coincidentally, similar concerns are also sometimes raised about the C++ language. These arguments have some merit, but on the margin more useful and actually *used* languages are preferable to those that are very cleanly designed, yet not used much.

Having a proper programming language is a key feature supporting rigorous and reproducible research: by encoding all steps of a data analysis and estimation in a script or program, the analyst makes every aspect of the process explicit and thereby ensures full reproducibility.

Consider the first example which is presented below. It is a slightly altered version of an example going back to a post by Greg Snow to the r-help mailing list.

```
xx <- faithful$eruptions
fit <- density(xx)
plot(fit)
```

Listing 1.1 Plotting a density in R

We assign a new variable xx by extracting the named component eruptions of the (two-column) data.frame faithful included with the R system. The data set contains waiting times between eruptions, as well as eruption duration times, at the Old Faithful geyser in the Yellowstone National Park in the USA. To estimate the density function of eruption duration based on this data, we then call the R function density (which uses default arguments besides the data we pass in). This function returns an object we named fit, and the plot function then visualizes it as shown in the corresponding Fig. 1.1.

This is a nice example, and it illustrates some features of R such as the object-oriented nature in which we can simply plot an object returned from a modeling function. However, this example was introduced primarily to provide the basis for an extension also provided by Greg Snow and shown in the next listing.

```
xx <- faithful$eruptions
fit1 <- density(xx)
fit2 <- replicate(10000, {
    x <- sample(xx,replace=TRUE);
    density(x, from=min(fit1$x), to=max(fit1$x))$y
})
fit3 <- apply(fit2, 1, quantile,c(0.025,0.975))
plot(fit1, ylim=range(fit3))
polygon(c(fit1$x,rev(fit1$x)),
        c(fit3[1,], rev(fit3[2,])),
        col='grey', border=F)
lines(fit1)
```

Listing 1.2 Plotting a density and bootstrapped confidence interval in R

1.1 Background: From R to C++

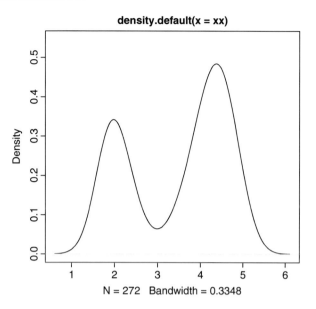

Fig. 1.1 Plotting a density in R

The first two lines are identical apart from now assigning to an object `fit1` holding the estimated density. Lines three to six execute a minimal bootstrapping exercise. The `replicate()` function repeats N (here 10,000) times the code supplied in the second argument. Here, this argument is a code block delimited by braces, containing two instructions. The first instruction creates a new data set by resampling with replacement from the original data. The second instruction then estimates a density on this resampled data. This time the data range is limited to the range of the initial estimated in `fit1`; this ensures that the bootstrapped density is estimated on the same grid of x values as in `fit1`. For this data set, the grid contains 512 points. We retain only the y coordinates of the fit—these will be collected as the N columns in the resulting object `fit2` making this a matrix of dimension $512 \times N$.

The next command on line 7 then applies the `quantile()` function to each of the 512 rows in `fit2`, returning the 2.5% and 97.5% quantiles, and creating a new matrix of dimension 2×512 where the two rows contains the quantile estimates at each grid point for the x axis. We then plot the initial fit, adjusting the y-axis to the range of quantile estimates. Next, we add a gray polygon defined by the x grid and the quantile estimates which visualizes the bootstrapped 95% confidence interval of the initial density estimate. Finally, we replot the `fit1` density over the gray polygon. The resulting plot is shown in Fig. 1.2.

The main takeaway of this second example is that with just a handful of lines of code, we can deploy fairly sophisticated statistical modeling functions (such as the density estimate) and even provide a complete resampling strategy. This uses the

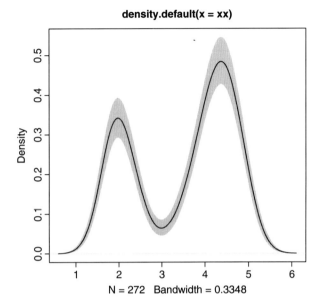

Fig. 1.2 Plotting a density and bootstrapped confidence interval in R

same estimation function in a nonparametric bootstrap to also provide a confidence interval for the estimation, and finally plots both. Few languages besides R are this expressive and powerful for working with data.

A key aspect of the internal implementation of R is that its own core interpreter and extension mechanism are implemented in the C language. C is often used for system programming as it is reasonably lean and fast, yet also very portable and easily available on most hardware platforms. A key advantage of C is that it is extensible via external libraries and modules. R takes full advantage of this, and so does the **Rcpp** extension featured in this book. The principal goal of **Rcpp** is to make writing these extensions easier and less error-prone. The aim of this book is to show how this can be accomplished with what we consider relative ease compared to the standard C interface.

The C language is also closely related to the C++ language which can be seen as an extension and superset. There are some small quibbles about a few minor aspects of C which do not carry over to C++ but we can safely ignore these for our purposes. C++ has been called "a federation of [four] languages" (Meyers 2005). This offers new and unique programming aspects and, in particular, provides a key match to the object-model in R (even if the terminology and philosophy of object-oriented programming differs between R and C++). Appendix A provides a very brief introduction to the C++ language.

1.2 A First Example

1.2.1 Problem Setting

Let us consider a concrete first example which requires only basic mathematics. This particular problem was first suggested in a post[1] at the StackOverflow site.

The Fibonacci sequence F_n is defined as a recursive sum of the two preceding terms in the same sequence:

$$F_n = F_{n-1} + F_{n-2} \qquad (1.1)$$

with these two initial conditions

$$F_0 = 0 \quad \text{and} \quad F_1 = 1$$

so that the first ten numbers of the sequence F_0 to F_9 are seen to be

$$0, 1, 1, 2, 3, 5, 8, 13, 21, 34.$$

Here, we follow the convention of starting the Fibonacci sequence at F_0; there are other discussions which begin the recursion with F_1 which would require slightly altered code in the examples shown below.

Fibonacci sequences have long been studied, and the corresponding Wikipedia page[2] provides additional resources. Fibonacci sequences can also be visualized: Fig. 1.3 shows the so-called *Fibonacci spiral* built from the first 34 Fibonacci numbers.

1.2.2 A First R Solution

The classic approach of implementing the computation of a Fibonacci number F_n for a given value of n is to evaluate equation 1.1 directly. This commonly leads to a simple recursive function.

In R this could be written as follows:

```
fibR <- function(n) {
    if (n == 0) return(0)
    if (n == 1) return(1)
    return (fibR(n - 1) + fibR(n - 2))
}
```

Listing 1.3 Fibonacci number in R via recursion

---

[1] See http://stackoverflow.com/questions/6807068/.
[2] See http://en.wikipedia.org/wiki/Fibonacci_number.

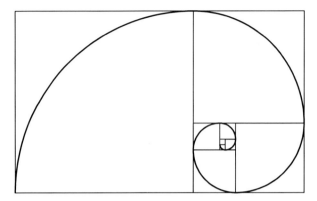

Fig. 1.3 Fibonacci spiral based on first 34 Fibonacci numbers

Source: http://en.wikipedia.org/wiki/File:Fibonacci_spiral_34.svg, released under Public Domain

This simple function has several key features:

- It is very short.
- It does not test for wrong input values less than zero.
- It is easy to comprehend.
- It is a very faithful rendition of the relationship in Eq. 1.1.

However, it also has a key disadvantage: it is very inefficient. Consider the calculation of F_5. Via the recursion in Eq. 1.1, this becomes the sum of F_3 and F_4. But already when we compute F_4 as the sum of F_3 and F_2, we note that we end up recomputing F_3. Similarly, F_2 will be computed several times too. In fact, formal analysis reveals that the algorithm is exponential in n—in other words its run-time increases at an exponential rate relative to the argument n. This type of performance is worst-in-class, leading to a search for alternative approaches.

Another concern specific to R is that function calls are not particularly lightweight, which makes recursive function calls particularly unattractive. Naturally, for both these reasons, many better algorithms have been suggested and we will discuss two other approaches below.

1.2.3 A First C++ Solution

A simple solution to compute F_n much faster using the same simple and intuitive algorithm is to switch to C or C++. We can write a simple C++ version as follows:

```
int fibonacci(const int x) {
    if (x == 0) return(0);
    if (x == 1) return(1);
    return (fibonacci(x - 1)) + fibonacci(x - 2);
}
```

Listing 1.4 Fibonacci number in C++ via recursion

1.2 A First Example

The function is recursive just like the preceding version. For simplicity, it also operates without checking its input arguments.

In order to call it from R, we need to use a wrapper function as R prescribes a very particular interface via its .Call() function: all variables used at the interface have to be of *pointer to S expression* type, or SEXP. There are alternatives to .Call(), but as we will discuss in Chap. 2, .Call() is our preferred interface as we can transfer whole objects from R to C++ and back which is not possible with the alternatives. So without going into details at this point, a suitable wrapper is

```
extern "C" SEXP fibWrapper(SEXP xs) {
    int x = Rcpp::as<int>(xs);
    int fib = fibonacci(x);
    return (Rcpp::wrap(fib));
}
```

Listing 1.5 Fibonacci wrapper in C++

This uses two key **Rcpp** tools, the converter functions as and wrap. The first, as, is used to convert the incoming argument xs from SEXP to integer. Similarly, wrap converts the integer result in the integer variable fib to the SEXP type returned by a function used with .Call().

1.2.4 Using Inline

The next steps are to compile these two functions, to link them into a so-called shared library (which can be loaded at run-time by a system such as R), and to actually load it. These three steps do sound a little tedious and labor-intensive, and they are. So it is at this point that we introduce another very powerful helper: the **inline** package (Sklyar et al. 2012).

inline, written mostly by Oleg Sklyar, brings an idea to R which has been used with other dynamically extensible scripting languages. By providing a complete wrapper around the *compilation*, *linking*, and *loading* steps, the programmer can concentrate on the actual code (in either one of the supported languages C, C++, or Fortran) and forget about the operating-system specific details of compilation, linking, and loading. A single entry point, the function cxxfunction() can be used to turn code supplied as a text variable into an executable function.

```
## we need a pure C/C++ function as the generated function
## will have a random identifier at the C++ level preventing
## us from direct recursive calls
incltxt <- '
int fibonacci(const int x) {
    if (x == 0) return(0);
    if (x == 1) return(1);
    return fibonacci(x - 1) + fibonacci(x - 2);
}'
```

```
## now use the snipped above as well as one argument conversion
## in as well as out to provide Fibonacci numbers via C++
fibRcpp <- cxxfunction(signature(xs="int"),
                      plugin="Rcpp",
                      incl=incltxt,
                      body='
    int x = Rcpp::as<int>(xs);
    return Rcpp::wrap( fibonacci(x) );
')
```

Listing 1.6 Fibonacci number in C++ via recursion, using inline

We actually supply two arguments, the pure C++ function and the wrapper function. The pure C++ function is passed as an argument to the `includes` argument which allows us to pass additional include directives, or even function or class definitions as seen here. This code passed to the `includes` variable is included unaltered in the code prepared by **inline**. The main body of the function is supplied as the argument to the argument `body`. The function `cxxfunction` writes the function signature (i.e., its interface defining the variables going in) using the information from the `signature` variable (here: a single argument named `xs`) and sets up the code to also use **Rcpp** features by selecting its plugin. We also note that the function body supplied here is identical to the wrapper functions detailed above.

Once `cxxfunction` has been run successfully, the resulting function created as a result (here: `fibRcpp()`) can be called just like any other R function.

In particular, we can time the execution and compare it to the first solution based on the initial R implementation. The results, obtained from running the script `fibonacci.r` included as an example in the **Rcpp** package, are shown in Table 1.1. The actual timing is undertaken by the package **rbenchmark** (Kusnierczyk 2012).

Table 1.1 Run-time performance of the recursive Fibonacci examples

Function	N	Elapsed time (s)	Relative (ratio)
fibRcpp	1	0.092	1.00
fibR	1	62.288	677.04
(byte-compiled) fibR	1	62.711	681.64

The compiled version is over 600 times faster, showing that recursive function calls do indeed exert a cost on R performance. We also notice that byte-compiling the R function does not make a difference as the results are essentially unchanged.

The takeaway from this section is that there is obvious merit in replacing simple R code with simple C++ code. Writing the Fibonacci recurrence as a simple three-line function is natural. Switching implementation languages to C++ very significantly boosts run-time performance as we have seen with certain values of the argument *n*. However, no matter what the chosen implementation language, an exponential algorithm will eventually be inapplicable provided the argument *n* is large enough.

For those cases, better algorithms help, and we will look at two different implementations below. It should, however, be stressed that faster implementation languages and better algorithms are not exclusive as we can combine both as we will do in the remainder of the chapter.

1.2.5 Using Rcpp Attributes

The **inline** package discussed in the previous section has become very widely used due to both its versatility and its robustness from fairly wide testing. As we have seen, it permits us to quickly extend R with compiled code directly from the R session.

More recently, **inline** has been complemented by a new approach that arrived with version 0.10.0 of the **Rcpp** package. This approach borrows from an upcoming (but not yet widely available) feature in the new C++ standard—the "attributes"— but implements it internally. The programmer simply declares certain "attributes," notably whether a function is to be exported for use from R or from another C++ function (or both). One can declare dependencies whose resolution still relies on the plugin framework provided by **inline**. Used this way, the "Rcpp attribute" framework can automate more aspects of the type conversion and marshaling of variables.

A simple example, once again on the Fibonacci sequence, follows:

```
#include <Rcpp.h>

using namespace Rcpp;

// [[Rcpp::export]]
int fibonacci(const int x) {
    if (x < 2)
        return x;
    else
        return (fibonacci(x - 1)) + fibonacci(x - 2);
}
```

Listing 1.7 Fibonacci number in C++ via recursion, using Rcpp attributes

The key element to note here is the `[[Rcpp::export]]` attribute preceding the function definition.

This can be used as easily as shown in the following example:

```
R> sourceCpp("fibonacci.cpp")
R> fibonacci(20)
[1] 6765
```

Listing 1.8 Fibonacci number in C++ via recursion, via Rcpp attributes and `sourceCpp`

The new function `sourceCpp()` reads the code from the given source file, parses it for the relevant attributes, and creates the required wrappers before calling R to compile and link just like **inline** does.

Note, however, that we did not have to specify a wrapper function, and that we obtain a single function `fibonacci()` which can operate through recursion.

The new "Rcpp attributes" will be discussed more in Sect. 2.6 where we revisit the same example using the `cppFunction()` which operates on a character string containing the program, rather than a file.

1.2.6 A Second R Solution

One elegant solution to retain the basic recursive structure of the algorithm without incurring the cost of repeated computation of the same value is provided by a method called *memoization*. Here, the Fibonacci number N is computed for each value between 1 and $N-1$ and stored. On the next computation, the precomputed value is recalled, as opposed to starting the full recursion. An R solution with memoization can be written as follows (and is courtesy of Pat Burns):

```
## memoization solution courtesy of Pat Burns
mfibR <- local({
    memo <- c(1, 1, rep(NA, 1000))
    f <- function(x) {
        if (x == 0) return(0)
        if (x < 0) return(NA)
        if (x > length(memo))
            stop("x too big for implementation")
        if (!is.na(memo[x])) return(memo[x])
        ans <- f(x-2) + f(x-1)
        memo[x] <<- ans
        ans
    }
})
```

Listing 1.9 Fibonacci number in R via memoization

If a value for argument *n* has already been encountered, it is used. Otherwise, it is computed and stored in vector memo. This ensures that the recursive function is called exactly once for each possible value of *n*, which results in a dramatic speedup.

1.2.7 A Second C++ Solution

We can also use memoization in C++. A simple solution is provided by the following piece of code:

```
## memoization using C++
mincltxt <- '
#include <algorithm>
```

1.2 A First Example

```
#include <vector>
#include <stdexcept>
#include <cmath>
#include <iostream>

class Fib {
public:
  Fib(unsigned int n = 1000) {
    memo.resize(n);                    // reserve n elements
    std::fill( memo.begin(), memo.end(), NAN ); // set to NaN
    memo[0] = 0.0;                     // initialize for
    memo[1] = 1.0;                     // n=0 and n=1
  }
  double fibonacci(int x) {
    if (x < 0)                         // guard against bad  input
      return( (double) NAN );
    if (x >= (int) memo.size())
      throw std::range_error(\"x too large for implementation\");
    if (! ::isnan(memo[x]))
      return(memo[x]);                 // if exist, reuse values
    // build precomputed value via recursion
    memo[x] = fibonacci(x-2) + fibonacci(x-1);
    return( memo[x] );                 // and return
  }
private:
  std::vector< double > memo;   // internal memory for precomp.
};
'
## now use the snippet above as well as one argument conversion
## in as well as out to provide Fibonacci numbers via C++
mfibRcpp <- cxxfunction(signature(xs="int"),
                        plugin="Rcpp",
                        includes=mincltxt,
                        body='
    int x = Rcpp::as<int>(xs);
    Fib f;
    return Rcpp::wrap( f.fibonacci(x-1) );
')
```

Listing 1.10 Fibonacci number in C++ via memoization

We define a very simple C++ class Fib with three elements:

- A constructor which is called once upon initialization.
- A single public member function which computes F_n.
- A private data vector holding the memoization values.

So this example provides a first glance at using classes in C++ code.

In the actual wrapper function, we simply instantiate an object f of the class Fib and then invoke the member function to compute the given Fibonacci number.

1.2.8 A Third R Solution

Naturally, we can also compute F_n using an iterative approach. A number of solutions are provided at the WikiBooks site[3] so that setting up a simple R solution such as the following is straightforward:

```r
## linear / iterative solution
fibRiter <- function(n) {
    first  <- 0
    second <- 1
    third  <- 0
    for (i in seq_len(n)) {
        third <- first + second
        first <- second
        second <- third
    }
    return(first)
}
```

Listing 1.11 Fibonacci number in R via iteration

The iterative solution improves further on the approach using memoization as it requires neither stateful memory nor recursion.

1.2.9 A Third C++ Solution

Given the iterative solution in R, it is also straightforward to write as a C++ function as shown below.

```
## linear / iterative solution
fibRcppIter <- cxxfunction(signature(xs="int"),
                           plugin="Rcpp",
                           body='
    int n = Rcpp::as<int>(xs);
    double first = 0;
    double second = 1;
    double third = 0;
    for (int i=0; i<n; i++) {
        third = first + second;
        first = second;
        second = third;
    }
    return Rcpp::wrap(first);
')
```

Listing 1.12 Fibonacci number in C++ via iteration

---

[3] http://en.wikibooks.org/wiki/Fibonacci_number_program.

For completeness, we also show a C++ solution that uses iterations. It is bound to be the fastest version yet as compiled loops generally execute faster than those from an interpreted language such as R.

1.3 A Second Example

1.3.1 Problem Setting

Let us consider a second example. This example was motivated in a private communication with Lance Bachmeier who used it in an introductory econometrics class. This example is also included in the **RcppArmadillo** package. **RcppArmadillo** uses **Rcpp** to implement a very convenient and powerful interface between R and the **Armadillo** library (Sanderson 2010) for linear algebra with C++. Chapter 10 provides a more in-depth discussion of **RcppArmadillo**.

The context of the example is a vector autoregressive process of order one for two variables, or in formal notation a VAR(1). More generally, a VAR model consists of a number K of endogenous variable $\mathbf{x}_t$. A VAR(p) process is then defined by a series of coefficient matrices A_j with $j \in 1,\ldots,p$ such that

$$\mathbf{x}_t = A_1 \mathbf{x}_{t-1} + \ldots + A_p \mathbf{x}_{t-p} + \mathbf{u}_t$$

plus a possible non-time-series regressor matrix which is omitted here. We follow typographic convention of using lowercase letters for scalars, bold lowercase letters for vectors, and uppercase letters for matrices.

For the example, we are considering the simplest case of a two-dimensional VAR of order one. At time t, it is comprised of two endogenous variables $\mathbf{x}_t = (x_{1t}, x_{2t})$ which are a function of their previous values at $t-1$ via a coefficient matrix A. As A is assumed to be constant, it no longer requires an index. This can be written as follows:

$$\mathbf{x}_t = A\mathbf{x}_{t-1} + \mathbf{u}_t \tag{1.2}$$

where $\mathbf{x}_t$ and $\mathbf{u}_t$ are time-varying vectors of size two and A is a two-by-two matrix.

1.3.2 R Solution

When studying the properties of VAR systems, simulation is a tool that is frequently used to assess these models. And, for the simulations, we need to generate suitable data. A closer look at Eq. 1.2 reveals that we cannot easily vectorize the expression due to the interdependence between the two coefficients. As a result, we need to loop explicitly.

```
## parameter and error terms used throughout
a <- matrix(c(0.5,0.1,0.1,0.5),nrow=2)
u <- matrix(rnorm(10000),ncol=2)

## Let's start with the R version
rSim <- function(coeff, errors) {
    simdata <- matrix(0, nrow(errors), ncol(errors))
    for (row in 2:nrow(errors)) {
        simdata[row,] = coeff %*% simdata[(row-1),] + errors[row,]
    }
    return(simdata)
}

rData <- rSim(a, u)                         # generated by R
```

Listing 1.13 VAR(1) of order 2 generation in R

This approach is pretty straightforward. The simulation function receives a 2×2 matrix a of parameters, and a vector u of size $N \times 2$ of normally and independently distributed random error terms. It then creates a vector y, also of dimension $N \times 2$, by looping from the second row to the last row. Each element of y is assigned the product of the previous row times the coefficient matrix plus the error terms as specified in equation 1.2.

1.3.3 C++ Solution

The same basic approach can be used with a C++ function to generate the simulated VAR data. The next listing shows how to use **RcppArmadillo** via **inline** to compile, link and load C++ code on the fly into your R session.

```
## Now load 'inline' to compile C++ code on the fly
suppressMessages(require(inline))
code <- '
    arma::mat coeff = Rcpp::as<arma::mat>(a);
    arma::mat errors = Rcpp::as<arma::mat>(u);
    int m = errors.n_rows;
    int n = errors.n_cols;
    arma::mat simdata(m,n);
    simdata.row(0) = arma::zeros<arma::mat>(1,n);
    for (int row=1; row<m; row++) {
        simdata.row(row) = simdata.row(row-1)*trans(coeff)
                         + errors.row(row);
    }
    return Rcpp::wrap(simdata);
'
## create the compiled function
```

1.3 A Second Example

```
18 rcppSim <- cxxfunction(signature(a="numeric",u="numeric"),
                          code,plugin="RcppArmadillo")
20
   rcppData <- rcppSim(a,u)                 # generated by C++ code
22
   stopifnot(all.equal(rData, rcppData))    # checking results
```
Listing 1.14 VAR(1) of order 2 generation in C++

We initialize a matrix for the coefficients and a matrix of error terms from the supplied function arguments. We then create a results matrix of the same dimension as the error term matrix, and loop as before to fill this result matrix row-by-row, just as we did in the preceding solution.

1.3.4 Comparison

We can run a comparison to determine the run-time of both approaches, as well as the run-time of a third hybrid solution using byte-compiled code (using the **compiler** package introduced with version 2.13.0 of the R system).

```
1 ## now load the rbenchmark package and compare all three
  suppressMessages(library(rbenchmark))
3 res <- benchmark(rcppSim(a,e),
                   rSim(a,e),
5                  compRsim(a,e),
                   columns=c("test", "replications", "elapsed",
7                            "relative", "user.self", "sys.self"),
                   order="relative")
```
Listing 1.15 Comparison of VAR(1) run-time between R and C++

The results, obtained from running the script varSimulation.r included as an example in the **RcppArmadillo**, are shown in Table 1.2.

Table 1.2 Run-time performance of the different VAR simulation implementations

Function	N	Elapsed (s)	Relative (ratio)
rcppSim	100	0.033	1.00
(byte-compiled) Rsim	100	2.229	67.55
rSim	100	4.256	128.97

The C++ solution takes only 33 ms whereas the R code takes 4.26 s, or almost 130 times as much. Byte-compilation improves the R performance by a factor of almost two—yet the byte-compiled function still trails the C++ solution by a factor of about 67.

1.4 Summary

This introductory chapter illustrated the appeal of using **Rcpp** to extend R with short and simple C++ routines. Code can be written in C++ which is very similar to the R code first used to prototype a solution. Thanks to tools such as the **inline** package and particularly the more recent Rcpp attributes, we can easily extend R with short C++ functions—and reap substantial performance gains in the process.

The rest of the book will introduce **Rcpp**, as well as extension packages such as **RcppArmadillo**, in much more detail. The next section starts with a fuller discussion of the required tools.

Chapter 2
Tools and Setup

Abstract Chapter 1 provided a gentle introduction to `Rcpp` and some of its key features. In this chapter, we look more closely at the required toolchain of compilers and related R packages needed to deploy the **Rcpp** package. In particular, on Windows, the Rtools collection is used and non-gcc compilers are not supported. On Unix-alike systems such as Linux and OS X, gcc/g++ is the default.

2.1 Overall Setup

The **Rcpp** package provides a `C++` Application Programming Interface (API) as an extension to the R system. Because of these very close ties to R itself, it is both bound by the choices made by the R build system and influenced by how R is configured.

Some of the requirements for working with **Rcpp** and R are:

- The development environment has to comprise a suitable compiler (which is discussed more in the next section), as well as header files and libraries for a number of required components (R Development Core Team 2012a).
- R should be built in a way that permits both dynamic linking and embedding; on Unix-alike systems this is typically ensured by the `--enable-shared-lib` option to `configure` (R Development Core Team 2012d, Chapter 8) and most binary distributions of R are built this way.
- Common development tools such as `make` are needed which should be standard on Unix-alike systems (though OS X requires installation of developer tools) whereas Windows users will have to install the Rtools suite provided via the CRAN mirror network (R Development Core Team 2012a, Appendix D).

In general, the standard environment for building a CRAN package from source is required. The (even stronger) requirement of being able to build R itself is a possible guideline as is documented in R Development Core Team (2012a,d).

There are a few additional CRAN packages that are very useful along with **Rcpp**, and which the package itself depends upon. These are:

inline which is invaluable for direct compilation, linking and loading of short code snippets, and used throughout this book too.

rbenchmark which is used to run simple timing comparisons and benchmarks; it is also recommended by **Rcpp** but not required.

RUnit which is used for unit testing; the package is recommended and will only be needed to rerun these tests but it is not strictly required.

We already saw two of these packages in use in the preceding chapter.

Lastly, users who want to build **Rcpp** from the repository source (rather than the distributed tarfile) also need the **highlight** binary by André Simon which is used to provide colored source code in several of the vignettes.

2.2 Compilers

2.2.1 General Setup

A basic requirement for extending a program with *suitable* loadable binary modules relates to the compiler being used. But exactly what is *suitable* can depend on a number of factors.

The choice of compilers generally matters, and more so for some languages than for others. The C language has a simpler interface for callable functions which makes it possible to have a program compiled with one compiler load a module built with another compiler. In general, this is not an option for C++ due to a much more complicated interface reflecting some of the richer structures in the C++ language. For example, how function names, and member function names inside classes, are represented is not standardized between compiler makers, and this generally prevents mixing of object code between different compilers.

As **Rcpp** is of course a C++ application, this last restriction applies and we need to stick with the compilers used to build R on the different platforms. The CRAN repository generally employs the same approach of using one main compiler per platform, and this approach is the one supported by the CRAN maintainers and the R Core team.

In practice, this means that on almost all platforms, the GNU Compiler Collection (or `gcc`, which is also the name of its C language compiler) has to be used along with the corresponding `g++` compiler for the C++ language. One notable exception is Solaris where the Sun compiler can be used as well; however, this platform is not as widely available and used, and we will not discuss its particular aspects any further. Also, on Windows, the prescribed way to access the suitable compiler is via the `Rtools` package contributed by the Windows R maintainers (R Development Core Team 2012a, Appendix D). OS X is an exception as Apple will not ship `gcc`

2.2 Compilers

versions past 4.2.1. Its transition to the `clang++` compiler of the LLVM project is not yet complete as this book is being written. Users on the OS X platform may have to download tools provided by Simon Urbanek, the R Core maintainer supporting OS X.

So on Windows, OS X and Linux, the compiler of choice for **Rcpp** generally is the `g++` compiler. A minimum suitable version is a final 4.2.* release; releases earlier than 4.2.* were lacking some C++ features used by **Rcpp**. Later versions are preferred as version 4.2.1 has some known bugs. But generally speaking, as of 2013 the (current) default compilers on all the common platforms are suitable. As of R version 2.12.0, the Windows platform has switched to version 4.5.1 of `g++` in order to support both 32- and 64-bit builds.

More advanced C++ features from the next C++ standard, C++11, which has recently been approved by the standards committee will become available once the compilers support them by default.

2.2.2 Platform-Specific Notes

Windows

Windows is both the most common platform for R use—yet quite possibly the hardest to develop on. The reason for this difficulty with R development on Windows is that the build environment and tools do not come standard with the operating system. However, due to the popularity of the platform, good support exists in the form of a third-party package kindly provided by some of the R Core developers who focus on Windows, namely Brian Ripley and Duncan Murdoch. The `Rtools` package, initially distributed via a site maintained by Duncan Murdoch but now available via the CRAN network, contains all the required tools in a single package. Complete instructions specific to Windows are available in the "R Administration" manual (R Development Core Team 2012a, Appendix D).

To stress again what was hinted at above: other compilers are not supported on Windows. In particular, the popular series of compilers produced by Microsoft cannot be used to build R from source (for reasons that are beyond the scope of this discussion) as these compilers are simply not supported by R Core. While it may be possible to compile some C++ extensions for R using these compilers, the **Rcpp** package follows the recommendation of the R Core team and sticks to the officially supported compilers. So for the last few years, and presumably for the next few years too, this limits the choice on Windows to the version of `g++` in the **Rtools** bundle.

OS X

OS X has become a popular choice of operating system among developers. As noted in the "R Administration" manual (R Development Core Team 2012a, Appendix C.4), the Apple Developer Tools (*e.g.*, Xcode) have to be installed (as well as gfortran if R or Fortran-using packages are to be built). Some older versions of OS X do not have a C++ compiler that is recent enough for some of the template code in the **Rcpp**; releases starting from "Snow Leopard" should be sufficient.

Unfortunately, Apple and the Free Software Foundation (the organization backing all the GNU software) are at an impasse over licensing. The GNU Compiler Collection now uses version 3 of GNU General Public License which Apple deems unsuitable for its operating system. As became cleat in 2011, it seems that g++ version 4.2.1 will be the last version available from Apple, which is unfortunate as more current g++ releases have made great strides towards adding new features of the upcoming C++ language standard. However, the clang++ compiler from the LLVM should eventually provide a full-featured replacement.

Linux

On Linux, developers need to install the standard development packages. Some distributions provide helper packages which pull in all the required packages; the r-base-dev package on Debian and Ubuntu is an example.

In general, whatever tools are needed to build R itself will be sufficient to build **Rcpp** from source, and to build packages utilizing **Rcpp**.

Other Platforms

Few other platforms appear to be in widespread use. The CRAN archive runs regression tests against Solaris and its Sun compiler. However, as we do not have direct access to the platform, development and debugging of **Rcpp** is somewhat cumbersome on this platform. Moreover, we have not yet detected measurable interest among the population of possible users. That said, **Rcpp** plugs into general R facilities for building packages, and the clear intent is to have **Rcpp** install and work on every platform supported by R itself.

2.3 The R Application Programming Interface

The R language and environment supports an application programming interface, or API for short. The API is described in the "Writing R Extensions" manual (R Development Core Team 2012d), and defined in the header files provided with every R installation. The R Core group usually stresses that only the public API should be used as other (undocumented) functions could change without notice.

Several books describe the API and its use. Venables and Ripley (2000) is an important early source. Gentleman (2009) and Matloff (2011) are more recent additions, while Chambers (2008) is authoritative in the context of "Programming with Data."

There are two fundamental extension functions provided: .C() and .Call(). The first, .C() first appeared in an earlier version of the R language and is much more restrictive. It only supports pointers to basic C types which is a very severe restriction. More current code uses the richer .Call() interface exclusively. It can operate on the so-called SEXP objects, which stands for pointers to S expression objects. Essentially everything inside R is represented as such a SEXP object, and by permitting exchange of such objects between the C and C++ languages on the one hand, and R on the other hand, programmers have the ability to operate directly on R objects. This is key for **Rcpp** as well—and the principal reason why **Rcpp** works exclusively with .Call().

Rcpp essentially sits on top of this API offered by R itself and provides a complementary interface to those aiming to extend R. By leveraging facilities available to C++ programmers (but not in plain C), **Rcpp** can offer what we think is an easier to use and possibly even more consistent interface that is closer to the way R programmers work with their data.

2.4 A First Compilation with Rcpp

Having discussed the required compiler and toolkit setup, and having seen introductory examples in Chap. 1, it is now appropriate to address how to use these tools on an actual source file. In doing so, we will use the explicit commands to illustrate the different steps required. Shorter and more convenient alternatives will be discussed later.

We consider the first example from the introductory chapter and assume that both the fibonacci function and the wrapper have been saved in a file fibonacci.cpp. Then, on a 64-bit Linux computer with **Rcpp** installed in a standard location, we can compile it via the example shown in Listing 2.1.

```
sh> PKG_CXXFLAGS="-I/usr/local/lib/R/site-library/Rcpp/include" \
    PKG_LIBS="-L/usr/local/lib/R/site-library/Rcpp/lib -lRcpp" \
    R CMD SHLIB fibonacci.cpp
g++ -I/usr/share/R/include -DNDEBUG \
    -I/usr/local/lib/R/site-library/Rcpp/include \
    -fpic -g -O3 -Wall -c fibonacci.cpp -o fibonacci.o
g++ -shared -o fibonacci.so fibonacci.o \
    -L/usr/local/lib/R/site-library/Rcpp/lib -lRcpp
    -Wl,-rpath,/usr/local/lib/R/site-library/Rcpp/lib \
    -L/usr/lib/R/lib -lR
```

Listing 2.1 A first manual compilation with **Rcpp**

Execution of `R CMD SHLIB` triggers two distinct g++ invocations. The first command (on line four) corresponds to `R CMD COMPILE` to turn a given source file into an object file. The second command (on line eight) corresponds to `R CMD LINK` and uses the g++ compiler a second time to link the object file into a shared library. This creates the file `fibonacci.so` which we can load into R. Also note how two environment variables are defined on lines 1 and 2 to let R know where to find the header files and libraries required for use with **Rcpp**.

But before we get to that step, let us review a few of the issues with the approach described here:

1. On line one, we have to set two environment variables, one each for the header file location (via `PKG_CXXFLAGS`) and one for the library location and name (via `PKG_LIBS`).
2. Both these variables use explicit path settings which are not portable across computers, let alone operating systems.
3. File extensions are operating-system dependent, the shared library ends on `.so` on Linux but `.dylib` under OS X.

To address some of these concerns, **Rcpp** offers two helper functions which can be invoked using the scripting front-end `Rscript`.

```
sh> PKG_CXXFLAGS=`Rscript -e 'Rcpp:::CxxFlags()'` \
    PKG_LIBS=`Rscript -e 'Rcpp:::LdFlags()'` \
    R CMD SHLIB fibonacci.cpp
```

Listing 2.2 A first manual compilation with **Rcpp** using Rscript

Running the example in Listing 2.2 results in the same two commands as above. But this approach improves over the previous one:

1. By using commands to request information which is returned in a portable fashion freeing the user from having to specify these details.
2. The helper functions are part of the **Rcpp** package and can therefore impute the relevant locations in a portable manner.
3. Moreover, the helper functions also know the operating system details and therefore are able to supply the required per-operating system details such as file extensions.

The end result is that we have a single command that works across platforms,[1] including portably in a `Makefile`. So we can now use this file in R.

```
R> dyn.load("fibonacci.so")
R> .Call("fibWrapper", 10)
[1]   55
```

Listing 2.3 Using the first manual compilation from R

---

[1] Well, Windows user may have to set the two environment variables differently but that is a shell limitation in Windows and not an issue with **Rcpp**.

We can load the shared library via the `dyn.load()` function. It uses the full filename, including the explicit platform-dependent extension which is .so on Unix, .dll on Windows, and .dylib on OS X. Once the shared library is loaded into the R session, we can then call the function `fibWrapper` using the standard `.Call()` interface. We supply the argument *n* to compute the corresponding Fibonacci number and obtain the requested result.

So this example proves the point we were trying to make in this section: we can extend R with simple C++ functions, even though the process of doing so may seem somewhat involved and intimidating at first. The **inline** package discussed in the next section and the Rcpp attributes extension discussed in the following section make the build process a lot more seamless to use.

2.5 The Inline Package

We saw in the previous chapter how to compile, link, and load a new function for use by R. We will now look more closely at a tool first mentioned in that introductory chapter which greatly simplifies this process.

2.5.1 Overview

Extending R with compiled code requires a mechanism for reliably compiling, linking, and loading the code. Doing this in the context of a package is preferable in the long run, but it may well be too involved for quick explorations. Undertaking the compilation manually is certainly possible. But, as the previous section showed, also somewhat laborious.

A better alternative is provided by the **inline** package (Sklyar et al. 2012) which compiles, links, and loads a C, C++ , or Fortran function—directly from the R prompt using simple functions `cfunction` and `cxxfunction`. The latter provides an extension which works particularly well with **Rcpp** via the so-called plug-ins which provide information about additional header file and library locations; and a third function, `rcpp`, which defaults to selecting that plugin for use with **Rcpp**.

The use of **inline** is possible as **Rcpp** itself can be installed and updated just like any other R package using, for example, the `install.packages()` function for initial installation as well as `update.packages()` for upgrades. So even though R / C++ interfacing would otherwise require source code, the **Rcpp** library is always provided ready for use as a pre-built library through the CRAN package mechanism.[2]

---

[2] This presumes a platform for which pre-built binaries are provided. **Rcpp** is available in binary form for Windows and OS X users from CRAN, and as a .deb package for Debian and Ubuntu users. For other systems, the **Rcpp** library is automatically built from source during installation or upgrades.

The library and header files provided by **Rcpp** for use by other packages are installed along with the **Rcpp** package. When building a package, the LinkingTo: Rcpp directive in the DESCRIPTION file lets R properly reference the header files automatically. That makes usage easier than for direct compilation via R CMD COMPILE or R CMD SHLIB (as in the previous section) where the function Rcpp:::CxxFlags() can be used to export the header file location and the appropriate -I switch. The **Rcpp** package also provides appropriate information for the -L switch needed for linking via the function Rcpp:::LdFlags(). It can be used by Makevars files of other packages, or to directly set the variables PKG_CXXFLAGS and PKG_LIBS, respectively.

The **inline** package makes use of both these facilities. All of this is done behind the scenes without the need for explicitly setting compiler or linker options. Moreover, by specifying the desired outcome rather to explicitly encode it, we provide a suitable level of indirection that permits the **Rcpp** package to completely abstract away the operating system-specific components. Usage of **Rcpp** via **inline** is therefore as portable as R itself: the same code will run on Windows, OS X, and Linux (provided the required tools are present as discussed earlier).

A standard example for a function extending R is a convolution of two vectors; this example is used throughout the "Writing R Extensions" manual (R Development Core Team 2012d). This convolution example can also be rewritten for use by **inline** as shown below. The function body is provided by the R character variable src, the function header (and its variables and their names) is defined by the argument signature, and we only need to enable plugin=="Rcpp" to obtain a new R function fun based on the C++ code in src:

```
R> src <- '
   Rcpp::NumericVector xa(a);
   Rcpp::NumericVector xb(b);
   int n_xa = xa.size(), n_xb = xb.size();

   Rcpp::NumericVector xab(n_xa + n_xb - 1);
   for (int i = 0; i < n_xa; i++)
      for (int j = 0; j < n_xb; j++)
         xab[i + j] += xa[i] * xb[j];
   return xab;
'
R> fun <- cxxfunction(signature(a="numeric", b="numeric"),
                     src, plugin="Rcpp")
R> fun( 1:4, 2:5 )
[1]  2  7 16 30 34 31 20
```

Listing 2.4 Convolution example using **inline**

With one assignment—albeit spanning lines one to eleven—to the R variable src, and one call of the R function cxxfunction (provided by the **inline** package), we have created a new R function fun that uses the C++ code we assigned to src—and all this functionality can be used directly from the R prompt making prototyping with C++ functions straightforward.

2.5 The Inline Package

Note that with version 0.3.10 or later of **inline**, a convenience wrapper `rcpp` is available which automatically adds the `plugin="Rcpp"` argument so that the invocation in Listing 2.4 could also have been written as

```
fun <- rcpp(signature(a="numeric", b="numeric"), src)
```

but we will generally use the `cxxfunction()` form.

A few further options are noteworthy at this stage. Adding `verbose=TRUE` shows both the temporary file created by `cxxfunction()` and the invocations by R CMD SHLIB. This can be useful for debugging if needed. Listing 2.5 shows the generated file. Noteworthy aspects include the function declaration with the randomized function name, and the signature with the two variable names implied from the `signature()` argument to `cxxfunction`. Also shown are the macros BEGIN_RCPP and END_RCPP discussed in Sect. 2.7.

Other options permit us to set additional compiler flags as well as additional include directories as shown in the next section.

2.5.2 Using Includes

As mentioned in the previous section, `cxxfunction` offers a number of other options. One aspect that we would like to focus on now is `includes`. As seen in Sect. 1.2.7, it allows us to include another block of code to, say, define a new `struct` or `class` type.

An example is provided by the following code sample from the Rcpp FAQ which was created after a user question on the **Rcpp** mailing list. A simple templated class which squares its argument is created in a code snippet supplied via `include`. The main function then uses this templated class on two different types:

```
1  R> inc <- '
   +     template <typename T>
3  +     class square : public std::unary_function<T,T> {
   +     public:
5  +         T operator()( T t) const { return t*t ;}
   +     };
7  '
   R> src <- '
9  +     double x = Rcpp::as<double>(xs);
   +     int i = Rcpp::as<int>(is);
11 +     square<double> sqdbl;
   +     square<int> sqint;
13 +     Rcpp::DataFrame df =
   +         Rcpp::DataFrame::create(Rcpp::Named("x", sqdbl(x)),
15 +                                 Rcpp::Named("i", sqint(i)));
   +     return df;
17 '
   R> fun <- cxxfunction(signature(xs="numeric", is="integer"),
19 +                     body=src, include=inc, plugin="Rcpp")
   R> fun(2.2, 3L)
```

```
>> Program source :

 1 :
 2 : // includes from the plugin
 3 :
 4 : #include <Rcpp.h>
 5 :
 6 :
 7 : #ifndef BEGIN_RCPP
 8 : #define BEGIN_RCPP
 9 : #endif
10 :
11 : #ifndef END_RCPP
12 : #define END_RCPP
13 : #endif
14 :
15 : using namespace Rcpp;
16 :
17 :
18 :
19 : // user includes
20 :
21 :
22 : // declarations
23 : extern "C" {
24 : SEXP file2370678f8cfe( SEXP a, SEXP b) ;
25 : }
26 :
27 : // definition
28 :
29 : SEXP file2370678f8cfe( SEXP a, SEXP b ){
30 : BEGIN_RCPP
31 :
32 :   Rcpp::NumericVector xa(a);
33 :   Rcpp::NumericVector xb(b);
34 :   int n_xa = xa.size(), n_xb = xb.size();
35 :   Rcpp::NumericVector xab(n_xa + n_xb - 1);
36 :   for (int i = 0; i < n_xa; i++)
37 :     for (int j = 0; j < n_xb; j++)
38 :       xab[i + j] += xa[i] * xb[j];
39 :   return xab;
40 :
41 : END_RCPP
42 : }
43 :
44 :
```

Listing 2.5 Program source from convolution example using **inline** in verbose mode

2.5 The Inline Package

```
      x i
1  4.84 9
```

Listing 2.6 Using **inline** with `include=`

This code example uses a few `Rcpp` items we have not yet encountered such as the `DataFrame` class or the static `create` method (and these will be discussed later). We again see the explicit converter `Rcpp::as<>()` used to access scalar types `integer` and `double` passed to C++ from R.

More important is the definition of the sample helper class `square`. It derives from a public class `std::unary_function` templated to the same argument and return type. It also defines just one `operator()` which, unsurprisingly for a class called `square`, returns its argument squared.

The example demonstrates that while `cxxfunction` may be of primary use for short and simple test applications, it can also be used to test in more complicated setups. In fact, the plugin structure discussed in the next section allows for even more customization, should it be needed. The **RcppArmadillo**, **RcppEigen**, and **RcppGSL** packages discussed in the final part of the book all use this facility via a plugin generator.

2.5.3 Using Plugins

We have seen the use of the options `plugin="Rcpp"` in the previous examples. Plugins provide a general mechanism for packages using **Rcpp** to supply additional information which may be needed to compile and link the particular package. Examples may include additional header files and directories, as well as additional library names to link against as well as their locations.

Without going into too much detail about how to write a plugin, we can easily illustrate the use of a plugin. Below is a example which shows the code underlying the `fastLm()` example from **RcppArmadillo**. We will rebuild it using `cxxfunction` from **inline**:

```
R> src <- '
+    Rcpp::NumericVector yr(ys);
+    Rcpp::NumericMatrix Xr(Xs);
+    int n = Xr.nrow(), k = Xr.ncol();
+
+    arma::mat     X(Xr.begin(), n, k, false);
+    arma::colvec  y(yr.begin(), yr.size(), false);
+
+    arma::colvec coef = arma::solve(X, y);    // fit y ~ X
+    arma::colvec res  = y - X*coef;           // residuals
+
+    double s2 = std::inner_product(res.begin(),res.end(),
+                                   res.begin(),double())
+                / (n - k);
+    arma::colvec se = arma::sqrt(s2 *
```

```
16  +              arma::diagvec(arma::inv(arma::trans(X)*X)));
    +
18  +   return Rcpp::List::create(Rcpp::Named("coef") = coef,
    +                             Rcpp::Named("se")   = se,
20  +                             Rcpp::Named("df")   = n-k);
    + '
22  R> fun <- cxxfunction(signature(ys="numeric", Xs="numeric"),
    +                     src, plugin="RcppArmadillo")
24  R> ## could now run fun(y, X) to regress y ~ X
```

Listing 2.7 A first **RcppArmadillo** example for **inline**

This illustrates nicely how **inline** can be used to compile, link, and load packages on the fly, even when these packages depend on several other R packages. In the case of **RcppArmadillo**, which integrates the Armadillo C++ library, the dependency is on both **RcppArmadillo** and **Rcpp**. The plugin provides the necessary information to compile and link this example.

2.5.4 Creating Plugins

A simple example of how to modify a plugin is provided in the Rcpp-FAQ vignette. This example is centered around using the GNU Scientific Library (or GNU GSL, or just GSL for short) along with R. The GSL is described in Galassi et al. (2010). The example here illustrates how to set a fixed header location. A more comprehensive example might also attempt to determine the location, possibly by querying the `gsl-config` helper script as done in the **RcppGSL** package discusses in Chap. 11.

```
    R> gslrng <- '
2   +   int seed = Rcpp::as<int>(par) ;
    +   gsl_rng_env_setup();
4   +   gsl_rng *r = gsl_rng_alloc (gsl_rng_default);
    +   gsl_rng_set (r, (unsigned long) seed);
6   +   double v = gsl_rng_get (r);
    +   gsl_rng_free(r);
8   +   return Rcpp::wrap(v);
    + '
10  R> plug <- Rcpp:::Rcpp.plugin.maker(
    +     include.before = "#include <gsl/gsl_rng.h>",
12  +     libs = paste("-L/usr/local/lib/R/site-library/Rcpp/lib "
    +                  "-lRcpp -Wl,-rpath,"
14  +                  "/usr/local/lib/R/site-library/Rcpp/lib ",
    +                  "-L/usr/lib -lgsl -lgslcblas -lm", sep=""))
16  R> registerPlugin("gslDemo", plug )
    R> fun <- cxxfunction(signature(par="numeric"),
18  +                     gslrng, plugin="gslDemo")
    R> fun(0)
20  [1] 4293858116
    R> fun(42)
22  [1] 1608637542
    R>
```

Listing 2.8 Creating a plugin for use with **inline**

Here the **Rcpp** function `Rcpp.plugin.maker` is used to create a plugin named `plug`. We specify the inclusion of the GSL header file declaring the random number generator functions. We also specify the required libraries for linking against the GSL (with values suitable for a Linux system). Subsequently, the plugin is registered and deployed in a call to `cxxfunction()`. Finally, we test the new function and generate two random draws for two different initial seeds.

2.6 Rcpp Attributes

A recent addition to **Rcpp** provides an even more direct connection between C++ and R. This feature is called "attributes" as it is inspired by a C++ extension of the same name in the new C++11 standard (which will be available to R users only when CRAN permits use of these extension, which may be years away).

Simply put, "Rcpp attributes" internalizes key features of the **inline** package while at the same time reusing some of the infrastructure built for use by **inline** such as the plugins.

"Rcpp attributes" adds new functions `sourceCpp` to source a C++ function (similar to how `source` is used for R code), `cppFunction` for a similar creation of a function from a character argument, `evalCpp` for a direct evaluation of a C++ expression and more.

Behind the scenes, these functions make use of the existing wrappers `as<>` and `wrap` and do in fact rely heavily on them: any arguments with existing converters to or from `SEXP` types can be used. The standard build commands such as `R RMD COMPILE` and `R CMD SHLIB` are executed behind the scenes, and template programming is used to provide compile-time bindings and conversion.

An example may illustrate this:

```
cpptxt <- '
int fibonacci(const int x) {
    if (x < 2) return(x);
    return (fibonacci(x - 1)) + fibonacci(x - 2);
}'

fibCpp <- cppFunction(cpptxt)       # compiles, load, links, ...
```

Listing 2.9 Example of new `cppFunction`

`cppFunction` returns an R function which calls a wrapper, also created by `cppFunction` in a temporary file which it also builds. The wrapper function in turn calls the C++ function we passed as a character string. The build process administered by `cppFunction` uses a caching mechanism which ensures that only one compilation is needed per session (as long as the source code used is unchanged).

Alternatively, we could pass the name of a file containing the code to the function `sourceCpp` which would compile, link, and load the corresponding C++ code and assign it to the R function on the left-hand side of the assignment.

These new attributes can also use **inline** plugins. The following simple example uses the plugin for the **RcppGSL** package (which is discussed more fully in Chap. 11). The program itself is not that interesting: we merely use the definitions of five physical constants.

```
R> code <- '
+ #include <gsl/gsl_const_mksa.h>       // decl of constants
+
+ std::vector<double> volumes() {
+     std::vector<double> v(5);
+     v[0] = GSL_CONST_MKSA_US_GALLON;       // 1 US gallon
+     v[1] = GSL_CONST_MKSA_CANADIAN_GALLON; // 1 Canadian gallon
+     v[2] = GSL_CONST_MKSA_UK_GALLON;       // 1 UK gallon
+     v[3] = GSL_CONST_MKSA_QUART;           // 1 quart
+     v[4] = GSL_CONST_MKSA_PINT;            // 1 pint
+     return v;
+ }'
R>
R> gslVolumes <- cppFunction(code, depends="RcppGSL")
R> gslVolumes()
[1] 0.003785412 0.004546090 0.004546092 0.000946353 0.000473176
R>
```

Listing 2.10 Example of new `cppFunction` with plugin

But as **inline** is very mature and tested, and as the attributes functions are at this point not of comparable maturity, the remainder of the book will continue to use **inline** and its slightly more verbose expression. Going forward more new documentation will probably be written using the new functions once the interface stabilizes. Transitioning from one system to the other is seamless as the examples above indicated.

2.7 Exception Handling

C++ has a mechanism for handling exceptions. At a conceptual level, this is similar to what R programmers may already be familiar with via the `tryCatch()` function, or its simpler version `try()`.

In essence, inside a segment of code preceded by the keyword `try`, an exception can be thrown via the keyword `throw` followed by an appropriately typed exception object which is typically inherited from the `std::exceptions` type.

The following example may illustrate this.

```
extern "C" SEXP fun( SEXP x ) {
    try {
        int dx = Rcpp::as<int>(x);
        if (dx > 10)
            throw std::range_error("too big");
        return Rcpp::wrap(dx * dx);
    } catch( std::exception& __ex__ ) {
```

2.7 Exception Handling

```
            forward_exception_to_r(__ex__);
9       } catch(...) {
            ::Rf_error( "c++ exception (unknown reason)" );
11      }
        return R_NilValue;   // not reached
13  }
```

Listing 2.11 C++ example of throwing and catching an exception

For reasons that will become apparent in a moment, we are showing a complete function rather than a just short snippet used with cxxfunction() from the **inline** package.

If this function is compiled and linked (with appropriate flags to find the **Rcpp** headers and library), we can call it as

```
1 R> .Call("fun", 4)
  [1] 16
3 R> .Call("fun", -4)
  [1] 16
5 R> .Call("fun", 11)
  Error in cpp_exception(message = "too big",
7     class = "std::range_error") : too big
  R>
```

Listing 2.12 Using C++ example of throwing and catching an exception

As the code tests only whether the argument is larger than 10, both 4 and −4 are properly squared by this (not very interesting) function. For the argument 11, however, the exception is triggered via the throw followed by exception of type std::range_error with a short text indicating that the argument is too large for the assumed parameter limitation.

What happens after the throw is that a suitable catch() segment is identified. Here, as the exception was typed with a type inherited from the standard exception, the first branch is the one the code enters. The exception is then passed to an internal **Rcpp** function which converts it into an R error message. And indeed, at the R level, we see both that an exception was caught and what its type was.

This is a very useful mechanism that permits the programmer to return control to the calling instance (here the R program) with a clearly defined message.

We can illustrate this last point with a second example. What happens when we call the function with a non-numeric argument?

```
  R> .Call("fun", "ABC")
2 Error in cpp_exception(message = "not compatible with INTSXP",
      class = "Rcpp::not_compatible") :
   not compatible with INTSXP
4 R>
```

Listing 2.13 C++ example of example from **Rcpp**-type checks

Here the function is called with a character variable which cannot be used in the assignment to the integer variable dx. So an exception is thrown by the templated **Rcpp** function as which is templated to an integer type (written as as<int>)

here. The exception that is thrown is of type `Rcpp::not_compatible` which also inherits from the standard exception and a proper R error message is generated. Similar messages will be shown if the **Rcpp** types discussed in the next two chapters are instantiated with inappropriate types.

If no matching type is found, the default `catch` branch is executed. Here, it simply calls the error function of the R API with a constant text message.

Because the framework of the `try` statement (preceding the actual code block) and the `catch` clauses at the end are in fact invariant, they can also be expressed as a simple unconditional macro. Such macros are provided by **Rcpp**. Their definitions are shown in Listing 2.14.

```
#ifndef BEGIN_RCPP
#define BEGIN_RCPP try{
#endif

#ifndef VOID_END_RCPP
#define VOID_END_RCPP } \
    catch (std::exception& __ex__) { \
       forward_exception_to_r( __ex__ ); \
    } \
    catch(...) { \
       ::Rf_error("c++ exception (unknown reason)"); \
    }
#endif

#ifndef END_RCPP
#define END_RCPP VOID_END_RCPP return R_NilValue;
#endif
```

Listing 2.14 C++ macros for **Rcpp** exception handling

These macros are also used by `cxxfunction()` so that the following function is fully equivalent to Listing 2.11.

```
src <- 'int dx = Rcpp::as<int>(x);
        if( dx > 10 )
            throw std::range_error("too big");
        return Rcpp::wrap( dx * dx);
')
fun <- cxxfunction(x="integer", body=src, plugin="Rcpp")
fun(3)
[1] 9
fun(13)
Error: too big
```

Listing 2.15 inline version of C++ example of throwing and catching an exception

Thanks to **inline**, this version is much easier to compile, link, and load. And of course, an Rcpp attributes version can be written just as easily:

```
cppFunction('
    int fun2(int dx) {
        if ( dx > 10 )
```

2.7 Exception Handling

```
4              throw std::range_error("too big");
         return dx * dx;
6     }
  ')
8 fun2(3)
  [1] 9
10 fun2(13)
  Error: too big
```

Listing 2.16 Rcpp attributes version of C++ example of throwing and catching an exception

The proper exception handling framework by **Rcpp** is provided automatically in both cases by adding the required code to the generated files.

Part II
Core Data Types

Chapter 3
Data Structures: Part One

Abstract This chapter first discusses the RObject class at the heart of the **Rcpp** class system. While RObject is not meant to be used directly, it provides the foundation upon which many important and frequently-used classes are built. We then introduce the two core vector types NumericVector and IntegerVector. Other related vector types are briefly discussed at the end of the chapter.

3.1 The RObject Class

The RObject class occupies a central role in the implementation of the **Rcpp** class hierarchy. While it is not directly user-facing, it provides the common structure used by all the classes detailed below. It is the basic class underlying the **Rcpp** API. We will therefore discuss aspects of this class before we turn to the key classes derived from it. An instance of the RObject class encapsulates an R object. Every R object itself is internally represented by a SEXP: a pointer to a so-called *S expression* object, or SEXPREC for short. The "R Internals" manual (R Development Core Team 2012b, Section 1.1) provides a fuller treatment of the SEXP pointers to SEXPREC types, as well as of the related VECSXP or vectors of *S expression* pointers. One key aspect is that *S expression* objects are union types (which are sometimes called variant types). This means that depending on the particular value in a control field, different types can be represented. One could think of this as being similar to a switch statement where, conditional on the value of the expression, one out of a set of given branches is executed. With a union type, depending on the value of the control field, the remaining bits will be interpreted as forming the type implied by the control field. Consequently, this implies that an object pointed to by a SEXP could, for example, hold an integer vector, whereas another object could hold a character string, or another type including one of the several internal types.

An important aspect of this representation is that SEXP objects should be considered *opaque*. In programming, this term usually refers to something which should be accessed and viewed only indirectly using helper functions. Such functions are provided by the R API for those using the C language. In particular, macros (in fact, two different sets of macros) are provided to access the SEXP types. Our **Rcpp** package extends this C language API with a higher-level abstraction provided via the C++ language.

Similar to the C API, the **Rcpp** API contains a set of consistent functions which are appropriate for all types. Key among these functions are those that manage memory allocation and de-allocation. As a general rule, users of the **Rcpp** API *never* need to manually allocate memory, or free it after use. This is an extremely important point as memory management has been vilified as a common source of programming mistakes. For example, the C language is described by critics as too error-prone due to the explicit and manual memory management it requires. Languages such as Java or C# aim to improve upon C by managing the memory for the users. C++ occupies a middle ground: programmers can manually control memory management (which is important for performance-critical application), yet language constructs such as the Standard Template Library (see Sect. A.5 for a brief introduction) also provide control structures (such as vectors and lists) which—by providing a higher-level abstraction—free the programmer from the manual and error-prone aspects of memory allocation and de-allocation. **Rcpp** follows this philosophy, and the RObject class is a key piece in the implementation as it transparently manages memory allocation and de-allocation for the users.

As for the actual implementation, the key aspect of the RObject class is that it is only a very thin layer around the SEXP type it encapsulates. This extends the *opaque* view approach of the R API by fully wrapping the underlying SEXP and providing the class member functions to access or modify it. One can think of the API provided by **Rcpp** as providing a richer, more complete means of accessing the underlying SEXP data representation which remains unaltered from the way R represents an object.

In fact, the SEXP is indeed the only data member of an RObject.[1] Hence, the RObject class does not interfere with the way R manages its memory. Nor does it not perform copies of the object into a different and possibly suboptimal C++ representation. Rather, it acts as a *proxy* to the object it encapsulates. With this, methods applied to the RObject instance are relayed back to the SEXP in terms of the standard R API functionality invoked by the proxy classes.

The RObject class also takes advantage of the explicit life cycle of C++ objects: objects which are dynamically allocated ("on the heap" in C or C++ parlance) using a so-called *constructor* member function are then automatically de-allocated at the end of local scope by the deconstructor. This lets the underlying R object (represented by the SEXP) be transparently managed by the R garbage collector. The RObject effectively treats its underlying SEXP as a resource. The constructor of the RObject class takes the necessary measures to guarantee that the underlying SEXP is protected from the garbage collector, and the destructor assumes the responsibility to withdraw that protection. Together, these two steps

---

[1] See the header file include/Rcpp/RObject.h for details.

provide transparent and automatic memory management for the user. And, by assuming the entire responsibility of garbage collection, **Rcpp** relieves the programmer from having to write repetitive code to manage the protection stack with the familiar PROTECT and UNPROTECT macros provided by the R API.

Aside from the memory management functionality, a number of helper functions are applicable to instances of the RObject class. As the underlying SEXP may be of different types, these member functions have to be applicable to any R object that can be represented by a SEXP, irrespective of its type.

Several member functions are available for all classes that are deriving from the RObject class. The isNULL, isObject, and isS4 functions are used to query object properties. The explicit naming of these functions provides a first description; these functions return a true or false value depending on the object they are being applied to. Similarly, the member function inherits can be used to test for inheritance from a specified class. Attributes of R objects can be queried or set with the functions attributeNames, hasAttribute, or attr. For S4 objects,[2] the hasSlot and slot functions permit the handling of the data slots that are a key feature of the S4 object system.

A large number of user-visible classes derive from the RObject class:

IntegerVector   for vectors of type integer.
NumericVector   for vectors of type numeric.
LogicalVector   for vectors of type logical.
CharacterVector   for vectors of type character.
GenericVector   for generic vectors which implement List types.
ExpressionVector   for vectors of expression types.
RawVector   for vectors of type raw.

For the integer and numeric types, we also have IntegerMatrix and NumericMatrix corresponding to the equivalent R types, and similarly implemented as vectors with associated dimension attributes specifying row and column sizes.

We will discuss the integer and numeric vectors in some detail— including examples—in the next two sections.

3.2 The IntegerVector Class

The IntegerVector class provides a natural mapping from and to the standard R integer vectors. We can assign existing R vectors to C++ objects, and we can create new integer vectors directly in C++ and return them to R. In both cases, the corresponding converter function—the templated as<>() function in the case of converting from R to C++ and the wrap() function for the inverse direction—are automatically called thanks to C++ template logic.

---

[2] Member functions dealing with slots are only applicable to S4 objects; otherwise an exception is thrown.

3.2.1 A First Example: Returning Perfect Numbers

Suppose we wanted to write a function that provides a vector with the first four perfect (and even) numbers. A perfect number is a positive integer that is equal to the sum of its divisors. The first one is six, as it is the sum of its divisors one, two, and three. The second perfect number is 28:

$$28 = 1 + 2 + 4 + 7 + 14$$

and two more even perfect numbers—496 and 8182—were already known by the ancient Greeks.[3]

With the help of the **inline** package introduced initially in Sect. 1.2.4 and more fully in Sect. 2.5, one can quickly create functions in R which are based on C++ code. The **inline** package takes the C++ source code as a character variable, and then compiles, links, and loads this code making it directly accessible via a function it returns. So here we assign five statements, separated by semicolons, to a single R character variable src. This variable is then passed to the cxxfunction() with additional arguments for a functions signature—empty in our case—and selection of the "Rcpp" plugin which will instruct cxxfunction() to look for the header files and library from the **Rcpp** package:

```
R> src <- '
+   Rcpp::IntegerVector epn(4);
+   epn[0] = 6;
+   epn[1] = 14;
+   epn[2] = 496;
+   epn[3] = 8182;
+   return epn;
+ '
R> fun <- cxxfunction(signature(), src, plugin="Rcpp")
R> fun()
[1]    6   14   496 8182
```

Listing 3.1 A function to return four perfect numbers

The example is of course not very meaningful—we could have created the same R vector in a single R statement. Yet the short program already highlights a few key points about the vector types:

- Creating a new vector is as easy as selecting an initial size (and there are other creation methods).
- Elements of the vector can be set one-by-one (and the new C++ standard C++11 will allow array-style assignments in one statement).
- Returning the vector requires no additional code as the implicit version of wrap is being called.

We will build upon this example in the next section.

---

[3] More details are at http://en.wikipedia.org/wiki/Perfect_number.

3.2.2 A Second Example: Using Inputs

The previous example showed how to create a new vector at the C++ level. Receiving a vector from R is also straightforward. Consider the next simple example which reimplements the prod() function for a given integer vector. Note that the use of the colon operator (:) creates as integer-valued sequence even though we do not use explicit integer instantiation via the L suffix (e.g., 10L).

```
 1 R> src <- '
   +     Rcpp::IntegerVector vec(vx);
 3 +     int prod = 1;
   +     for (int i=0; i<vec.size(); i++) {
 5 +         prod *= vec[i];
   +     }
 7 +     return Rcpp::wrap(prod);
   '
 9 R> fun <- cxxfunction(signature(vx="integer"), src,
   +                    plugin="Rcpp")
11 R> fun(1:10)   # creates integer vector
   [1] 3628800
```

Listing 3.2 A function to reimplement prod()

The example shows a second possibility for instantiation of an IntegerVector object. In this case, through implicit use of the as<>() template function, the SEXP-typed argument vx is used. This variable is defined in the function signature via the first argument to cxxfunction. Through vx, the vector vec has been instantiated: It contains a copy of the pointer to the original SEXP object from R. It is important to stress that only the pointer to the underlying data is copied, not the underlying data itself.

And given vec, it is straightforward to compute the product. We can also solve this problem using tools from the Standard Template Library (STL) as shown in the next example.

```
   R> src <- '
 2 +     Rcpp::IntegerVector vec(vx);
   +     int prod = std::accumulate(vec.begin(),vec.end(),
 4 +                                1, std::multiplies<int>());
   +     return Rcpp::wrap(prod);
 6 + '
   R> fun <- cxxfunction(signature(vx="integer"), src,
 8 +                    plugin="Rcpp")
   R> fun(1:10)   # creates integer vector
10 [1] 3628800
```

Listing 3.3 A second function to reimplement prod()

This approach employs the accumulate() function, which is in the std namespace like the rest of the STL. It is called with iterators—which we can think of as functions which safely generalize the notion of a pointer—to the beginning and the end of the vector. This allows the function to operate on the thereby chosen range of elements. The next two arguments are an initial value of one, just like in the previous example, and a binary function, in this case the predefined function `multiplies` which is templated to the integer type to correspond to our vector type. Usage of STL may appear a little more complicated at first. But just how functional programming in R tends to become more natural with use, extended use of the STL is certainly a recommended programming habit for C++.

These two code examples, though simple to understand and hence suitable for exposition, still have few flaws we would not use in more serious code. First, no test is made for `NA` or zero values in the vector. Second, the code is likely to do poorly on larger values due to integer overflow: even the sequence `1L:13L` returns a result that is already different from what `prod` returns, so switching to computing the product as a sum of logarithms may be preferable, while possibly being more expensive to compute (yet this could be provided as an option). Third, and on a more cosmetic note, we could even have omitted the assignment to the temporary variable and returned the result from `accumulate` directly in `wrap`.

3.2.3 A Third Example: Using Wrong Inputs

An important feature of the class hierarchy is the ability to test for conforming input types. Consider a function written for an integer vector (as above). What behavior shall we expect with types that are different? This could be a floating-point vector where an integer vector is expected: automatic conversion would be nice. But what behavior would be expected for clearly inappropriate types?

It so happens that we can illustrate this behavior using some of the example programs shown above. Let us revisit the `prod`-replacement expecting an `integer` vector.

```
R> fun(1:10)
[1] 3628800
R> fun(seq(1.0, 1.9, by=0.1))
[1] 1
```

Listing 3.4 Testing the `prod()` function with floating-point inputs

The first result restates what we had seen before: floating-point number (which happen to be whole numbers) can be converted without a problem to the corresponding integers. The second example is more interesting: The product of ten floating-point number over the interval from 1.0 to 1.9 yields 1? How?

The answers lie in the typical behavior of computers when confronted with floating-point and integer numbers. Here, we assigned a vector of floating-point numbers—all greater or equal to one—to an integer vector. Standard behavior in this case is *truncation* (rather than rounding). So the value 1.5 simply becomes 1. And consequently, the *integer* product of ten floating-point values between 1.0 and 1.9 is indeed 1 as the calculation reduces to 1^{10}.

But what happens when we use clearly inappropriate types?

```
R> fun(LETTERS[1:10])
Error in fun(LETTERS[1:10]) : not compatible with INTSXP
```

Listing 3.5 Testing the `prod()` function with inappropriate inputs

An exception is thrown when the integer vector object is instantiated as no conversion from character to integer is possible. This exception is caught and then transformed into an R error message, while control (in the interactive session) resumes—as we discussed above in Sect. 2.7. In other words, the type-conversion code behind the **Rcpp** object hierarchy both tests for appropriate types and safely returns control to the R session in case an inadmissible type is used as input.

Last but not least, we note that R integer vectors can be converted as easily into `std::vector<int>`. Similarly, the `NumericVector` type discussed in the next section can be converted in `std::vector<double>`.

3.3 The NumericVector Class

3.3.1 A First Example: Using Two Inputs

`NumericVector` is quite possibly the most commonly used vector type among the **Rcpp** classes. It corresponds to the basic R type of a numeric vector and can hold real-valued floating-point variables. Its storage type is `double`, and all computation will be in double precision just as in R itself.

As a first example, consider a simple generalization of a sum of squares calculation. Instead of always squaring the elements, we also pass an argument for the exponent.

```
R> src <- '
+     Rcpp::NumericVector vec(vx);
+     double p = Rcpp::as<double>(dd);
+     double sum = 0.0;
+     for (int i=0; i<vec.size(); i++) {
+         sum += pow(vec[i], p);
+     }
+     return Rcpp::wrap(sum);
+ '
R> fun <- cxxfunction(signature(vx="numeric", dd="numeric"),
+                     src, plugin="Rcpp")
R> fun(1:4,2)
[1] 30
R> fun(1:4,2.2)
[1] 37.9185
```

Listing 3.6 A function to return a generalized sum of powers

This example could also be rewritten using an STL algorithm. But using a custom-written transformation function would be a little more involved due to its C++ focus and detract us from examining the C++ and R integration.

3.3.2 A Second Example: Introducing `clone`

One important aspect of the proxy model implementation mentioned above is that the C++ object contains a pointer to the underlying SEXP object from R, which is itself a pointer. This implies that code trying to transform a vector—say by taking logarithms—and wanting to return a modified copy along with the original vector cannot be written as follows where both vectors are constructed from the input argument:

```
R> src <- '
+   Rcpp::NumericVector invec(vx);
+   Rcpp::NumericVector outvec(vx);
+   for (int i=0; i<invec.size(); i++) {
+     outvec[i] = log(invec[i]);
+   }
+   return outvec;
+ '
R> fun <- cxxfunction(signature(vx="numeric"),
+                     src, plugin="Rcpp")
R> x <- seq(1.0, 3.0, by=1)
R> cbind(x, fun(x))
               x
[1,] 0.0000000 0.0000000
[2,] 0.6931472 0.6931472
[3,] 1.0986123 1.0986123
```

Listing 3.7 Declaring two vectors from the same SEXP type

Modifications in outvec do, due to its underlying pointer sharing with the same underlying R object, also affect invec. Changes will therefore also affect the R object passed in as an argument. So while this lightweight proxy model makes for efficient code, we need different operations to create an independent second vector. The clone method is a suitable alternative as it allocates memory for a new object. Hence, changes do not propagate to the original vector:

```
R> src <- '
+   Rcpp::NumericVector invec(vx);
+   Rcpp::NumericVector outvec = Rcpp::clone(vx);
+   for (int i=0; i<invec.size(); i++) {
+     outvec[i] = log(invec[i]);
+   }
+   return outvec;
+ '
R> fun <- cxxfunction(signature(vx="numeric"),
+                     src, plugin="Rcpp")
R> x <- seq(1.0, 3.0, by=1)
R> cbind(x, fun(x))
       x
[1,] 1 0.0000000
[2,] 2 0.6931472
[3,] 3 1.0986123
R>
```

Listing 3.8 Declaring two vectors from the same SEXP type using clone

3.3 The NumericVector Class

We should note that `clone` is a generic feature of vectors derived from `RObject` objects and applies to all objects instantiated from a `SEXP`.

To close this point, we should also note that an even simpler form uses Rcpp sugar (discussed more fully in Chap. 8) to directly assign the result via a single vectorized call of the `log()` function:

```
 1 R> src <- '
   +     Rcpp::NumericVector invec(vx);
 3 +     Rcpp::NumericVector outvec = log(invec);
   +     return outvec;
 5 + '
   R> fun <- cxxfunction(signature(vx="numeric"),
 7 +                      src, plugin="Rcpp")
   R> x <- seq(1.0, 3.0, by=1)
 9 R> cbind(x, fun(x))
          x
11 [1,]   1 0.0000000
   [2,]   2 0.6931472
13 [3,]   3 1.0986123
   R>
```

Listing 3.9 Using Rcpp sugar to compute a second vector

We could even have omitted the declaration of, and assignment to, `outvec` and computed the result directly in the `return` statement thanks to the implicit use of `wrap()`.

3.3.3 A Third Example: Matrices

Besides vectors, matrices play an equally important role in modeling, as they do in the underlying linear algebra derivations. Internally, matrices are implemented as vectors with an associated dimension attribute, just as they are in R itself. Similarly, the more general form is actually a multidimensional array—and the matrix is merely a special case where the dimension attribute has size two for both a row and column count.

For example, a numeric vector of dimension three can be created as

```
    Rcpp::NumericVector vec3 =
 2      Rcpp::NumericVector( Rcpp::Dimension(4, 5, 6));
```

Listing 3.10 Declaring a three-dimensional vector

These multidimensional arrays can be useful for particular applications. However, we will focus more on matrices for their more common use in linear algebra and modeling.

A first example illustrates the use of matrices and also shows the `clone` method discussed in the previous section.

```
R> src <- '
2 +     Rcpp::NumericMatrix mat =
  +         Rcpp::clone<Rcpp::NumericMatrix>(mx);
4 +     std::transform(mat.begin(), mat.end(),
  +         mat.begin(), ::sqrt);
6 +     return mat;
  + '
8 R> fun <- cxxfunction(signature(mx="numeric"), src,
  +                    plugin="Rcpp")
10 R> orig <- matrix(1:9, 3, 3)
R> fun(orig)
12         [,1]    [,2]    [,3]
[1,] 1.00000 2.00000 2.64575
14 [2,] 1.41421 2.23607 2.82843
[3,] 1.73205 2.44949 3.00000
16 R>
```

Listing 3.11 A function to take square roots of matrix elements

This also illustrates how the two-dimensional matrix is treated as a one-dimensional continuous vector (just like in R) in memory as the sqrt() function is being swept across all elements.

Many of the "Rcpp sugar" extensions discussed more fully in Chap. 8 are also directly applicable to vectors and matrices.

3.4 Other Vector Classes

3.4.1 LogicalVector

The LogicalVector class is very similar in behavior to IntegerVector as it represents the two possible values of a logical, or boolean, type. These values—True and False—can also be mapped to one and zero (or more generally to "not zero" and zero).

However, as R generally supports missing values in its data structures, the LogicalVector has to support this—and is, in fact, seen as supporting three rather than two possible values. Listing 3.12 illustrates this as it shows how the other nonfinite values NaN, Inf, and NA all collapse into NA in the context of a logical vector.

```
R> fun <- cxxfunction(signature(), plugin="Rcpp",
2 +                   body='
  +     Rcpp::LogicalVector v(6);
4 +     v[0] = v[1] = false;
  +     v[1] = true;
6 +     v[3] = R_NaN;
  +     v[4] = R_PosInf;
8 +     v[5] = NA_REAL;
```

3.4 Other Vector Classes

```
   +        return v;
10 + ')
   R> fun()
12 [1] FALSE    TRUE FALSE       NA       NA       NA
   R> identical(fun(), c(FALSE, TRUE, FALSE, rep(NA, 3)))
14 [1] TRUE
   R>
```

Listing 3.12 A function to assign a logical vector

The example shows that assigning any of three possible nonfinite values NaN, Inf (which can be positive or negative), or NA (which is commonly defined for real-valued variables only) results in NA value in the logical vector.

The example also illustrates, by means of using the R function identical, that the values returned from an **Rcpp**-created function are indistinguishable from those created directly in R itself.

3.4.2 CharacterVector

The class CharacterVector can be used for vectors of R character vectors ("strings").

```
 1 R> fun <- cxxfunction(signature(), plugin="Rcpp",
   +                    body='
 3 +     Rcpp::CharacterVector v(3);
   +     v[0] = "The quick brown";
 5 +     v[1] = "fox";
   +     v[2] = R_NaString;
 7 +     return v;
   + ')
 9 R> fun()
   [1] "The quick brown" "fox"                     NA
11 R>
```

Listing 3.13 A function to assign a character vector

And similar to the other vectors, CharacterVector can hold its primary type, here strings, as well as NA value. Character vectors can also be converted to std::vector<std::string>.

3.4.3 RawVector

The RawVector can be very useful when "raw" bytes have to be used, for example in a networked application that transmits them to another application or program running on another machine. As such a use case is somewhat more specialized, we are not showing a full example here.

Chapter 4
Data Structures: Part Two

Abstract This chapter introduces several other important classes such as `List`, `DataFrame`, `Function`, and `Environment` which both correspond to key R language objects and have an underlying `SEXP`-representation.

The previous chapter discussed fundamental **Rcpp** classes centered around the basic `Vector` classes, covering anything from integers and numeric to raw types, logicals vector and string vectors. It also touched upon multidimensional vectors and the important special case of matrices.

In this chapter, we extend the analysis to more data types. After introducing a useful helper class, we continue with what seems like yet another vector, but is really the all-important `List` type. This is followed by the `DataFrame` class, before we turn to different types of classes with `Function` and `Environment`, respectively. We then briefly discuss `S4` and `Reference Classes` before ending the chapter with the R mathematics functions.

4.1 The Named Class

The `Named` class is a helper class used for setting the key side of key/value pairs. It corresponds to standard R use. Consider this simple example

```
1 R> someVec <- c(mean=1.23, dim=42.0, cnt=12)
  R> someVec
3 mean    dim    cnt
  1.23  42.00  12.00
5 R>
```

Listing 4.1 A named vector in R

where three elements are assigned in a vector, and each is also assigned a specific identifier or label. The `Named` class permits us to do the same thing for objects created in C++ which we want to return to the calling R function with such labels.

Several examples of this class will follow in the next sections. As a first illustration, consider the vector shown above which could be created at the C++ level as follows:

```
1 R> src <- '
  +     Rcpp::NumericVector x =
3 +         Rcpp::NumericVector::create(
  +             Rcpp::Named("mean") = 1.23,
5 +             Rcpp::Named("dim")  = 42,
  +             Rcpp::Named("cnt")  = 12);
7 +     return x; '
  R> fun <- cxxfunction(signature(), src, plugin="Rcpp")
9 R> fun()
    mean   dim   cnt
11  1.23 42.00 12.00
```

Listing 4.2 A named vector in C++

We can shorten the somewhat verbose coding style by

- Declaring using namespace Rcpp; to import the **Rcpp** namespace (and we should note that the cxxfunction() function from the **inline** package also does that when the "Rcpp" plugin is selected)
- Employing the shortcut form _["key"]

which allows us to rewrite the example as

```
1 R> src <- '
  +     NumericVector x = NumericVector::create(
3 +             _["mean"] = 1.23,
  +             _["dim"]  = 42,
5 +             _["cnt"]  = 12);
  +     return x; '
7 R> fun <- cxxfunction(signature(), src, plugin="Rcpp")
  R> fun()
9   mean   dim   cnt
    1.23 42.00 12.00
```

Listing 4.3 A named vector in C++, second approach

We may switch between the more explicit and the shortened style for some of the following examples.

4.2 The List aka GenericVector Class

The GenericVector is equivalent to the List type. This is the most general data type which can contain other types—and corresponds to how a list() in R can contain objects of different types. Objects of type List can also contain other objects of different lengths (and, this is sometimes called a "ragged array" if vectors of different length are parts of the same object). Furthermore, List objects can contain other List objects, which allow for arbitrary nesting of data structures.

4.2 The List aka GenericVector Class

Being able to hold different types of objects makes the `List` type suitable for parameter exchanges in either direction between R and C++.

4.2.1 List to Retrieve Parameters from R

Consider the following example, taken from the general-purpose optimization package **RcppDE** (Eddelbuettel 2012b). It has been somewhat simplified by removing a number of similar argument types, and by removing a second layer dealing with exception handling.

```
RcppExport SEXP DEoptim(SEXP lowerS, SEXP upperS,
                        SEXP fnS, SEXP controlS, SEXP rhoS) {

    Rcpp::NumericVector  f_lower(lowerS), f_upper(upperS);
    Rcpp::List           control(controlS);
    double VTR              = Rcpp::as<double>(control["VTR"]);
    int i_strategy          = Rcpp::as<int>(control["strategy"]);
    int i_itermax           = Rcpp::as<int>(control["itermax"]);
    int i_D                 = Rcpp::as<int>(control["npar"]);
    int i_NP                = Rcpp::as<int>(control["NP"]);
    int i_storepopfrom      = Rcpp::as<int>(control["storepopfrom"])-1;
    int i_storepopfreq      = Rcpp::as<int>(control["storepopfreq"]);
    int i_specinitialpop    = Rcpp::as<int>(control["specinitialpop"]);
    Rcpp::NumericMatrix initialpopm =
               Rcpp::as<Rcpp::NumericMatrix>(control["initialpop"]);
    double f_weight         = Rcpp::as<double>(control["F"]);
    double f_cross          = Rcpp::as<double>(control["CR"]);
    [...]
}
```

Listing 4.4 Using the `List` class for parameters

Here two vectors for upper and lower parameter bounds are directly initialized from one `SEXP` each as we have seen in the previous chapter. The `SEXP` variable `controlS` is assigned to a `Rcpp::List` variable named `control`. This object contains a large set of user-supplied parameters to control the optimization.

The `List` type allows for access by named string "key," similar to how we would extract a named entry from a list in R using the `[["key"]]` operator. Here in C++, we obtain an element of the list which will generally be of `SEXP` type—and we can use the explicit conversion function `as<>()` along with a template type to assign the value.

For example, the first parameter is a floating-point variable keyed to the name "VTR" which we assign to a `double`. Similarly, several count variables denoting sizes, dimensions, numbers of iterations, etc. are assigned to more integer-valued variable. However, the list also contains a genuine numeric matrix under the key "initialpop." **RcppDE** performs differential optimization, an evolutionary algorithm related to genetic algorithms but particularly suitable for floating-point representations. These types of optimization algorithms operate on populations of potential solutions—and the matrix indexed by "initialpop" can be used to initialize the algorithm with an initial set of potential solutions. Finally, two more floating-point control parameters are assigned.

4.2.2 List to Return Parameters to R

We can use an example from the same package, **RcppDE**, to illustrate how to return values to C++ as well.

```
1  return Rcpp::List::create(Rcpp::Named("bestmem")    = t_bestP,
                             Rcpp::Named("bestval")    = t_bestC,
3                            Rcpp::Named("nfeval")     = l_nfeval,
                             Rcpp::Named("iter")       = i_iter,
5                            Rcpp::Named("bestmemit")  =
                                            t(d_bestmemit),
7                            Rcpp::Named("bestvalit")  = d_bestvalit,
                             Rcpp::Named("pop")        = t(d_pop),
9                            Rcpp::Named("storepop")   = d_storepop);
```

Listing 4.5 Using a `List` to return objects to R

Because we do not show the declaration of the variables, we cannot tell immediately what their types are. But the beauty of the `List` type is that all types that can be converted to a `SEXP` are admissible! Here we have **Armadillo** vectors (`t_bestP`) and matrices (`d_bestmemit, d_pop`) thanks to **RcppArmadillo**, as well as standard `long` (`_nfeval`), `double` (`t_bestC`) and `int` (`i_iter`) scalars.

The use of `Rcpp::List::create()` is fairly idiomatic. It permits us to create a list on the fly, with its size determined at compile-time by the number of `name = value` pairs we have supplied. However, as we have seen in earlier examples, an alternate method of setting elements is also available by first reserving a sufficiently dimensioned list (and the same is of course true for vectors). This can be done directly using the constructor as in `Rcpp::List ll(4)`, where four elements would be reserved. A second possibility is to use the `reserve()` member function to specify a size. Once sufficient space has been reserved, we can then assign these using the standard square bracket operator `[]`. Needless to say, the square bracket operator cannot assign elements beyond the pre-reserved size range.

An alternative insertion method is provided by two functions modeled after equivalent STL functions: `push_back()` which insert the given element at the back—and, thereby, extends the vector or list by one element—and `push_front()` which inserts at the front and similarly extends the size by one. It should, however, be noted that these may alter the vectors. And as vectors are implemented in contiguous memory, this will in most cases result in a complete copy of the whole vector. In other words, the operations `push_back` and `push_front` operations can be relatively inefficient (due to the underlying memory model of `SEXP` types) and are provided mainly for convenience.

4.3 The DataFrame Class

Data frames are an essential object type in R and are used by almost all modeling functions, so naturally **Rcpp** supports this type too. Internally, data frames are represented as lists. This permits the data frame to contain data of different types. For example, a data frame may contain time stamps, real-valued measurements as well as, say, group identifications encoded as a factor. The different columns will always be *recycled* to have common length. To take an example, if we insert a vector of length four into a data frame followed by a vector of length two, the latter vector will be repeated a second time to also have length four (and this recycling at construction is done only for integer multiples). Having common length is an important feature, as other functions can always assume that data frames are rectangular. Rows are commonly seen as observations with columns representing variables.

We have already seen one example of a data frame creation earlier in Sect. 2.5.2 where the static `create` function was used. A similar example, taken from one of the unit tests, is

```
R> src <- '
    Rcpp::IntegerVector v =
                Rcpp::IntegerVector::create(7,8,9);
    std::vector<std::string> s(3);
    s[0] = "x";
    s[1] = "y";
    s[2] = "z";
    return Rcpp::DataFrame::create(Rcpp::Named("a")=v,
                                   Rcpp::Named("b")=s);
'
R> fun <- cxxfunction(signature(), src, plugin="Rcpp")
R> fun()
   a b
1  7 x
2  8 y
3  9 z
```

Listing 4.6 Using the `DataFrame` class

Otherwise, the data frame type can really be seen as a specialization of the list type, with the added restrictions of excluding nesting types and of imposing common length. While the latter is achieved via recycling in R, in C++ we have to ensure that each component of a `data.frame` is of the same length.

Data frames are very useful to concisely organize return data for further use by R, as well as a standard data representation for many modeling functions.

4.4 The Function Class

4.4.1 A First Example: Using a Supplied Function

A function object is needed whenever an R function—either supplied by the user or by accessing an R function—is employed. Consider this simple first example which uses the `sort()` function (passed as an argument from R) and applies it to a user-supplied vector:

```
R> src <- '
2 +     Function sort(x) ;
  +     return sort( y, Named("decreasing", true));
4 + '
R> fun <- cxxfunction(signature(x="function",
6 +                             y="ANY"),
  +                   src, plugin="Rcpp")
8 R> fun(sort, sample(1:5, 10, TRUE))
  [1] 5 5 5 3 3 3 2 2 2 1
10 R> fun(sort, sample(LETTERS[1:5], 10, TRUE))
   [1] "E" "E" "C" "B" "B" "B" "B" "B" "A" "A"
```

Listing 4.7 Using a `Function` passed as argument

On line two a C++ variable named `sort` is initialized from the object x that is of type `function`. The object named y is passed through as is; we never instantiate a C++ object with it. After compiling and loading this function, we pass the R function `sort()` as the first argument. Other suitable functions such as `order()` could also be used.

Another useful point to note is that because the second argument is never instantiated, we can pass different types of a suitable nature. In the example above, both integer and character vectors are passed in as randomized permutations, and both are being returned in decreasing sort order. This works because no **Rcpp** object is instantiated and hence no particular type is encoded (or even enforced via static typing). So no tests for matching types are executed and no exceptions are thrown—as was the case in Sect. 3.2.3—because no mismatched type is encountered. This example shows that passing the original SEXP type through can have its uses too.

4.4.2 A Second Example: Accessing an R Function

The `function` class can also be used to access R functions directly. In the example below, we draw five random numbers from a t-distribution with three degrees of freedom. As we are accessing the random number generators, we need to ensure that it is in a proper state. The `RNGScope` class ensures this by initializing the random number generator by calling the `GetRNGState()` function from the class constructor, and by restoring the initial state via `PutRNGState()` via its destructor (R Development Core Team 2012d, Section 6.3).

4.5 The Environment Class

```
R> src <- '
+      RNGScope scp;
+      Rcpp::Function rt("rt");
+      return rt(5, 3);
+  '
R> fun <- cxxfunction(signature(),
+                     src, plugin="Rcpp")
R> set.seed(42)
R> fun()
[1]   2.339681   0.130995  -0.074028  -0.057701  -0.046482
R> fun()
[1]   9.16504    1.08153    0.87017    1.99557   -0.22438
```

Listing 4.8 Using a Function accessed from R

We first instantiate a function object by giving it a string with the name of the R function we want to access. If the function is not globally accessible, we may need to access the corresponding namespace first.

Regarding the random number generation, it is also important to note that executing the equivalent commands in R itself, that is, set.seed(42) followed by rt(10,3) to generate ten random numbers from the *t*-distribution with three degrees of freedom, generates exactly the same ten numbers. This is key for reproducibility and also helps with debugging.

4.5 The Environment Class

The environment provides access to environments, a type of object which may be familiar to R programmers. Environments are defined in Section 2.1.10 of the "R Language" manual (R Development Core Team 2012c). Their role in variable lookup and their relationship to namespaces are described in Section 1.2 of the "R Internals" manual (R Development Core Team 2012b).

As a first example of using an environment, we consider the following example where we instantiated the stats namespace of R to access the rnorm() function:

```
    Rcpp::Environment stats("package:stats");
    Rcpp::Function rnorm = stats["rnorm"];
    return rnorm(10, Rcpp::Named("sd", 100.0));
```

Listing 4.9 Using a Function via an Environment

As the preceding section showed, such a two-step approach may not be needed as we can also use the Function class in **Rcpp** to search for an identifier.

However, it is useful to create and initialize environments, or to use environments to access variables in the current R session. A second example interfaces the global environment in R:

```
Rcpp::Environment global =
    Rcpp::Environment::global_env();
std::vector<double> vx = global["x"];

std::map<std::string,std::string> map;
map["foo"] = "oof";
map["bar"] = "rab";

global["y"] = map;
```

Listing 4.10 Assigning in the global environment

Here an instance of the global environment is created using the variable name global. It is used to access a variable x in R via direct lookup. Similarly, we create a map from string to string of size two and assign it to a symbol y in the global environment.

4.6 The S4 Class

R as a programming language has evolved over several decades. The first big step towards object-oriented programming was provided by what is known as the "White Book" (Chambers and Hastie 1992) which introduced S3 classes and methods. While particular to R and the underlying S programming language, and thus different from object-oriented programming notions in C++ or Java, S3 methods provide a simple approach that is still supported and is widely used. The basic feature is called method dispatch, implemented via what is known as a "generic function" which invokes corresponding methods driven by the class of the data type. A useful introduction to this approach is provided by Venables and Ripley (2000, Chapter 4).

A more ambitious approach to object-oriented programming was then added with the introduction of S4 classes in the "Green Book" (Chambers 1998); a more recent treatment is provided by Chambers (2008, Chapters 9 and 10). These classes have also been in R for well over a decade and provide a significant extension to the preceding object-oriented programming framework available to the R programmer. S4 classes offer a rich structure with a more rigid formalism, at the cost of some of the flexibility of the earlier and somewhat more ad hoc S3 class types. However, S4 offers more structure which may be needed for larger programming tasks.

As indicated towards the end of Sect. 3.1, **Rcpp** can access and modify S4 objects using the Rcpp::S4 class type. Getting and setting slot types of S4 objects is supported, as are various tests for object properties such S4 objects.

Listing 4.11 shows how to test if RObject object is in fact an S4 object, how to test for presence of a slot, and how to access a slot.

```
f1 <- cxxfunction(signature(x="any"), plugin="Rcpp", body='
    RObject y(x) ;
    List res(3) ;
    res[0] = y.isS4();
```

```
5   res[1] = y.hasSlot("z");
    res[2] = y.slot("z");
7   return res;
    ')
```

Listing 4.11 A simple example for accessing S4 class elements

Similarly, Listing 4.12 shows how to create an S4 object at the C++ level

```
    f2 <- cxxfunction(signature(x="any"), plugin="Rcpp", body='
2   S4 foo(x);
    foo.slot(".Data") = "foooo";
4   return foo;
    ')
```

Listing 4.12 A simple example for accessing S4 class elements

So while the basic facilities to access, alter, or create S4 objects exist, it may often remain easier to do so at the R code level. So a more common paradigm may be to compute and create the core C++ aspects of an object at the C++ level and to complement the objects at the R language level. Given that **Rcpp** functionality is often accessed from R functions, additional R code can then be executed at the R level after returning from a C++ function.

S4 classes have been used extensively in a number of CRAN packages. The BioConductor Project uses them throughout a large number of its packages as well.

4.7 ReferenceClasses

ReferenceClasses appeared with R version 2.12.0 and complete the set of object-oriented programming paradigms in R by adding a style more similar to what is found in C++ or Java. As of late 2012, the best documentation for ReferenceClasses is provided by invoking `help(ReferenceClasses)` in R; the topic is still undergoing changes.

ReferenceClasses are implemented using S4 methods and classes and are therefore related to S4 at least in the sense of some of the implementation details. ReferenceClasses are also related to "Rcpp modules" (covered later in Chap. 7) which use them as a representation.

Two key aspects of ReferenceClasses are (a) that they are mutable which differs from the standard "copy on write" semantics in the R system and (b) that methods are primarily associated with objects rather than functions.

Their mutable state makes ReferenceClasses objects good candidates for use in applications that require to track a "state." Typical examples are graphical user interface applications or servers. With the mutable state also comes the pass-by-reference semantics: rather than copying the whole object (when a component changes), only the reference to the object is copied. This is closer to how C++ and Java objects behave. Similarly, the association of methods to the underlying objects also more closely resembles the object-oriented design philosophies of C++ and Java.

ReferenceClasses are still undergoing changes and more definitive documentation may be forthcoming. Because this area of R programming has not yet settled as firmly, we will limit the discussion of **Rcpp** in this context to this brief overview.

4.8 The R Mathematics Library Functions

R provides a large number of mathematical and statistical functions described in the header file Rmath.h. As documented in R Development Core Team (2012d, Section 6.16), these functions can be used from a stand-alone library independent of R itself. Naturally, they can also be used with the R API.

R programmers may want to access these functions from C++ code too. **Rcpp** provides these via the "Rcpp sugar" extension described below in Chap. 8 as vectorized functions. In order to use them on atomistic double types, the prefix Rf_ was required. Starting with **Rcpp** release 0.10.0, these functions are also accessible via the R namespace. By using a distinct namespace, it is possible to cleanly reuse the same identifiers without the need for a remapping prefix such as Rf_ even though the functions are provided only via a more limited C language API.

The following example illustrates this use. For a given vector X, the probability function of the Normal distribution is computed.

```
#include <Rcpp.h>

extern "C" SEXP mypnorm(SEXP xx) {
    Rcpp::NumericVector x(xx);
    int n = x.size();
    Rcpp::NumericVector y1(n),y2(n),y3(n);

    for (int i=0; i<n; i++) {
        // accessing function via remapped R header
        y1[i] = ::Rf_pnorm5(x[i], 0.0, 1.0, 1, 0);

        // or accessing same function via Rcpp's 'namespace R'
        y2[i] = R::pnorm(x[i], 0.0, 1.0, 1, 0);
    }
    // or using Rcpp sugar which is vectorized
    y3 = Rcpp::pnorm(x);

    return Rcpp::DataFrame::create(Rcpp::Named("R")=y1,
                                    Rcpp::Named("Rf_")=y2,
                                    Rcpp::Named("sugar")=y3);
}
```

Listing 4.13 Example use of Rmath.h functions

A vector x is instantiated, its size determined, and three return vectors are allocated. Each will contain the corresponding pnorm value. On line nine, the new form R::pnorm() is used. It is identical to the function described in the documentation

4.8 The R Mathematics Library Functions

for the R API, but provided from a distinct namespace R. Line ten shows the older approach: a global identifier (as indicated by ::) with the prefix Rf_ which is effectively a primitive C language attempt at creating a namespace separation when no such feature exists in the language. Both functions operate on a single double-type variable at a time; a loop is required to compute all elements of the vector.

However, line twelve provides a preview of "Rcpp sugar" as detailed in Chap. 8. This version is in fact vectorized and all components of the vector y3 are assignment in a single (vectorized) operation just as one would in R but at the speed of C++.

The R namespace contains a large number of probability, density, and quantile functions, as well as corresponding random numbers, for a wide variety of distributions: Normal, Uniform, Gamma, Beta, LogNormal, Chi-squared, F, t, Binomial, Multinomial, Cauchy, Exponential, Geometric, Hypergeometric, Negative Binomial, Poisson, Weibull, Non-central Beta, Non-central F, Non-central t, Studentized Range (also known as "Tukey"), Wilcoxon Rank Sum, Wilcoxon Signed Rank, as well as a number of related functions.

The header file Rmath.h within the **Rcpp** header directory provides full details.

Part III
Advanced Topics

Chapter 5
Using Rcpp in Your Package

Abstract This chapter provides an overview of how to use **Rcpp** when writing an R package. It shows how using the function Rcpp.package.skeleton() can create a complete and self-sufficient example of a package using **Rcpp**. All components of the directory tree created by Rcpp.package.skeleton() are discussed in detail. A brief case study of an existing CRAN package concludes the chapter.

This chapter complements the *Writing R Extensions* manual (R Development Core Team 2012d) which is the authoritative source on how to extend R in general.

5.1 Introduction

Rcpp helps to extend R by offering an easy-to-use yet featureful interface between C++ and R. **Rcpp** itself is distributed as an R package. However, it is somewhat different from a traditional R package because its key component is a C++ library along with a set of header files defining the library interface. A client package that wants to make use of the **Rcpp** features must link against this library provided by **Rcpp**. For most of the previously examined examples, we have relied on the **inline** package to take care of the finer details of the dependencies on the **Rcpp** package itself.

It should be noted that R has only limited support for C and C++-level dependencies between packages (R Development Core Team 2012d). The LinkingTo declaration in the package DESCRIPTION file allows the client package to access the headers of the target package (here **Rcpp**), but support for linking against a library is not provided by R (outside of a function registration setup more suitable for C-only packages with a limited number of registered interface functions) and has to be added manually.

The discussion in this chapter follows the Rcpp.package.skeleton() function to show a recommended way of using **Rcpp** from a client package. We illustrate this using a simple C++ function which will be called by an R function. The material in the *Writing R Extensions* manual (R Development Core Team 2012d) is strongly recommended. Other documents on R package creation (for example Leisch 2008) are helpful as well concerning the package build process. R package creation is standardized and follows a logical pattern—but is not all that well documented leading beginners to experience some frustration. A basic understanding of how to create an R package helps when trying to add the additional information needed on how to use the **Rcpp** package in such add-on packages.

The working example provided in the next few sections provides a complete illustration of the process and can serve as a reference use case.

5.2 Using Rcpp.package.skeleton

5.2.1 Overview

Rcpp provides a function Rcpp.package.skeleton, modeled after the base R function package.skeleton, which facilitates creation of a so-called skeleton package using **Rcpp**. A skeleton package is a minimal package providing a working example which can then be adapted and extended as needed by the user.

Rcpp.package.skeleton has a number of arguments documented on its help page (and similar to those of package.skeleton). The main argument is the first one which provides the name of the package one aims to create by invoking the function. An illustration of a call using the argument mypackage is provided below.

```
R> Rcpp.package.skeleton( "mypackage" )
Creating directories ...
Creating DESCRIPTION ...
Creating NAMESPACE ...
Creating Read-and-delete-me ...
Saving functions and data ...
Making help files ...
Done.
Further steps are described in
    './mypackage/Read-and-delete-me'.

Adding Rcpp settings
 >> added Depends: Rcpp
 >> added LinkingTo: Rcpp
 >> added useDynLib directive to NAMESPACE
 >> added Makevars file with Rcpp settings
 >> added Makevars.win file with Rcpp settings
 >> added example header file using Rcpp classes
 >> added example src file using Rcpp classes
 >> added example R file calling the C++ example
 >> added Rd file for rcpp_hello_world
R>
```

Listing 5.1 A first Rcpp.package.skeleton example

5.2 Using Rcpp.package.skeleton

We can use the (Linux) command `ls -1R` to recursively list the directory and file structure created by this command:

```
R> system("ls -1R mypackage")
mypackage:
DESCRIPTION
man
NAMESPACE
R
Read-and-delete-me
src

mypackage/man:
mypackage-package.Rd
rcpp_hello_world.Rd

mypackage/R:
rcpp_hello_world.R


mypackage/src:
Makevars
Makevars.win
rcpp_hello_world.cpp
rcpp_hello_world.h
R>
```

Listing 5.2 Files created by Rcpp.package.skeleton

Using Rcpp.package.skeleton() is by far the simplest approach as it fulfills two roles. It creates the complete set of files needed for a package, and it also includes the different components needed for using **Rcpp** that we discuss in the following sections.

5.2.2 R Code

The skeleton created here contains an example R function rcpp_hello_world() that uses the .Call interface to invoke the C++ function rcpp_hello_world from the package **mypackage**.

```
rcpp_hello_world <- function(){
    .Call( "rcpp_hello_world", PACKAGE = "mypackage" )
}
```

Listing 5.3 R function rcpp_hello_world

Rcpp uses the .Call calling convention as it allows exchange of actual R objects back and forth between the R side and the C++ side. R objects encoded as SEXP types can be conveniently manipulated using the **Rcpp** API as we discussed in the preceding chapters.

Note that in this example, no arguments are passed from R down to the C++ layer. Doing so is straightforward (and one of the key features of **Rcpp**) but not central to our discussion of the package creation mechanics and hence omitted here.

5.2.3 C++ Code

The C++ function is declared in the rcpp_hello_world.h header file:

```
#ifndef _mypackage_RCPP_HELLO_WORLD_H
#define _mypackage_RCPP_HELLO_WORLD_H

#include <Rcpp.h>

/*
 * note : RcppExport is an alias to `extern "C"`
 * defined by Rcpp.
 *
 * It gives C calling convention to the rcpp_hello_world
 * function so that it can be called from .Call in R.
 * Otherwise, the C++ compiler mangles the
 * name of the function and .Call can't find it.
 *
 * It is only useful to use RcppExport when the function
 * is intended to be called by .Call. See the thread
 * http://thread.gmane.org/gmane.comp.lang.r.rcpp/649/focus=672
 * on Rcpp-devel for a misuse of RcppExport
 */
RcppExport SEXP rcpp_hello_world();

#endif
```

Listing 5.4 C++ header file rcpp_hello_world.h

The header includes the Rcpp.h file which is the sole header file that needs to be included in order to use **Rcpp**. The function itself is implemented in the file rcpp_hello_world.cpp.

```
#include "rcpp_hello_world.h"

SEXP rcpp_hello_world(){
    using namespace Rcpp ;

    CharacterVector x = CharacterVector::create( "foo", "bar" );
    NumericVector y   = NumericVector::create( 0.0, 1.0 );
    List z            = List::create( x, y );

    return z ;
}
```

Listing 5.5 C++ source file rcpp_hello_world.cpp

5.2 Using Rcpp.package.skeleton

The function creates an R list that contains a character vector and a numeric vector using **Rcpp** classes. At the R level, we will therefore receive a list of length two containing these two vectors:

```
R> rcpp_hello_world( )
[[1]]
[1] "foo" "bar"
[[2]]
[1] 0 1
R>
```

Listing 5.6 Calling R function rcpp_hello_world

5.2.4 DESCRIPTION

The skeleton generates an appropriate DESCRIPTION file, using both Depends: and LinkingTo: for **Rcpp**:

```
Package: mypackage
Type: Package
Title: What the package does (short line)
Version: 1.0
Date: 2012-11-10
Author: Who wrote it
Maintainer: Who to complain to <yourfault@somewhere.net>
Description: More about what it does (maybe more than
  one line)
License: What Licence is it under ?
LazyLoad: yes
Depends: Rcpp (>= 0.9.13)
LinkingTo: Rcpp
```

Listing 5.7 DESCRIPTION file for skeleton package

Rcpp.package.skeleton() adds the three last lines to the DESCRIPTION file. The Depends declaration indicates R-level dependency between the client package and **Rcpp**. The LinkingTo declaration indicates that the client package needs to use header files exposed by **Rcpp**.

5.2.5 Makevars and Makevars.win

Unfortunately, and notwithstanding its name, the LinkingTo declaration in itself is not enough to link to the user C++ library of **Rcpp**. Until more explicit support for libraries is added to R, we need to manually add the **Rcpp** library to the PKG_LIBS variable in the Makevars and Makevars.win files. **Rcpp** provides the unexported function Rcpp:::LdFlags() to ease the process:

```
 1 ## Use the R_HOME indirection to support
   ## installations of multiple R version
 3 PKG_LIBS=`$(R_HOME)/bin/Rscript -e "Rcpp:::LdFlags()"`

 5 ## As an alternative, one can also add this code in a
   ## file 'configure'
 7 ##
   ##   PKG_LIBS=`${R_HOME}/bin/Rscript -e
 9 ##                       "Rcpp:::LdFlags()"`
   ##
11 ##   sed -e "s|@PKG_LIBS@|${PKG_LIBS}|" \
   ##      src/Makevars.in > src/Makevars
13 ##
   ## which together with the following file
15 ## 'src/Makevars.in'
   ##
17 ##   PKG_LIBS = @PKG_LIBS@
   ##
19 ## can be used to create src/Makevars dynamically. This
   ## scheme is more powerful and can be expanded to also
21 ## check for and link with other libraries. It should
   ## be complemented by a file 'cleanup'
23 ##
   ##   rm src/Makevars
25 ##
   ## which removes the autogenerated file src/Makevars.
27 ##
   ## Of course, autoconf can also be used to write
29 ## configure files. This is done by a number of
   ## packages, but recommended only for more advanced
31 ## users comfortable with autoconf and its related
   ## tools.
```

Listing 5.8 Makevars file for skeleton package

The file Makevars.win is the equivalent version targeting Windows. This version uses an additional variable to call the architecture-dependent variant of Rscript in order to create the correct arguments for 32-bit or 64-bit versions of Windows.

```
  ## Use the R_HOME indirection to support
2 ## installations of multiple R version
  PKG_LIBS = $(shell
4    "${R_HOME}/bin${R_ARCH_BIN}/Rscript.exe"
     -e "Rcpp:::LdFlags()")
```

Listing 5.9 Makevars.win file for skeleton package

5.2.6 NAMESPACE

The `Rcpp.package.skeleton()` function also creates a file NAMESPACE with the content shown here.

```
1  useDynLib(mypackage)
   exportPattern("^[[:alpha:]]+")
```

Listing 5.10 NAMESPACE file for skeleton package

This file serves two purposes. First, it ensures that the dynamic library contained in the package we are creating via `Rcpp.package.skeleton()` will be loaded and thereby made available to the newly created R package. Second, it declares which identifiers, that is functions or data sets, should be globally visible from the namespace of this package. As a reasonable default, we export all functions via a regular expression covering all identifiers starting with a letter.

5.2.7 Help Files

Also created is a directory `man` containing two help files. One is for the package itself, the other for the (single) R function being provided and exported. Writing a help file is an important step in fully documenting a package. The *Writing R Extensions* manual (R Development Core Team 2012d) provides the complete documentation on how to create suitable content for help files.

5.2.7.1 mypackage-package.Rd

The help file `mypackage-package.Rd` is used to describe the new package.

```
   \name{mypackage-package}
2  \alias{mypackage-package}
   \alias{mypackage}
4  \docType{package}
   \title{
6    What the package does (short line)
   }
8  \description{
     More about what it does (maybe more than one line)
10   ~~ A concise (1-5 lines) description of the package ~~
   }
12 \details{
     \tabular{ll}{
14     Package: \tab mypackage\cr
       Type: \tab Package\cr
16     Version: \tab 1.0\cr
       Date: \tab 2012-11-10\cr
18     License: \tab What license is it under?\cr
```

```
         LazyLoad: \tab yes\cr
20   }
     ~~ An overview of how to use the package, including
22   the most important functions ~~
     }
24   \author{
       Who wrote it
26
       Maintainer: Who to complain to <yourfault@somewhere.net>
28   }
     \references{
30   ~~ Literature or other references for background
     information ~~
32   }
     ~~ Optionally other standard keywords, one per line,
34   from file KEYWORDS in the R documentation directory ~~
     \keyword{ package }
36   \seealso{
       ~~ Optional links to other man pages, e.g. ~~
38     ~~ \code{\link[<pkg>:<pkg>-package]{<pkg>}} ~~
     }
40   \examples{
       %% ~~simple examples of the most important functions~~
42   }
```

Listing 5.11 Manual page mypackage-package.Rd for skeleton package

5.2.7.2 rcpp_hello_world.Rd

The help file rcpp_hello_world.Rd serves as documentation for the example R function.

```
    \name{rcpp_hello_world}
2   \alias{rcpp_hello_world}
    \docType{package}
4   \title{
      Simple function using Rcpp
6   }
    \description{
8     Simple function using Rcpp
    }
10  \usage{
      rcpp_hello_world()
12  }
    \examples{
14    \dontrun{
        rcpp_hello_world()
16    }
    }
```

Listing 5.12 Manual page rcpp_hello_world.Rd for skeleton package

5.3 Case Study: The wordcloud Package

An interesting package which uses **Rcpp** in what may be the simplest possible way is **wordcloud** (Fellows 2012).

The **wordcloud** package has one main function creating word clouds, a common illustration depicting relative word frequency in a text corpus. The package initially provided this functionality via an R solution. However, the performance of iteratively finding placements of words on a two-dimensional plane such that the placement is tight yet not overlapping was seen as limiting. So the key determination of whether there is overlap between boxes, which executes a loop over the potentially large list of key words assigned to boxes, was then reimplemented in a short C++ function which is shown in Listing 5.13.

```
#include "Rcpp.h"

/*
 * Detect if a box at position (x11,y11), with width sw11
 * and height sh11 overlaps with any of the boxes in boxes1
 */
using namespace Rcpp;

RcppExport SEXP is_overlap(SEXP x11,SEXP y11,SEXP sw11,
                           SEXP sh11,SEXP boxes1){
    double x1 = as<double>(x11);
    double y1 =as<double>(y11);
    double sw1 = as<double>(sw11);
    double sh1 = as<double>(sh11);
    Rcpp::List boxes(boxes1);
    Rcpp::NumericVector bnds;
    double x2, y2, sw2, sh2;
    bool overlap= true;
    for (int i=0;i < boxes.size();i++) {
        bnds = boxes(i);
        x2 = bnds(0);
        y2 = bnds(1);
        sw2 = bnds(2);
        sh2 = bnds(3);
        if (x1 < x2)
            overlap = (x1 + sw1) > x2;
        else
            overlap = (x2 + sw2) > x1;

        if (y1 < y2)
            overlap = (overlap && ((y1 + sh1) > y2));
        else
            overlap = (overlap && ((y2 + sh2) > y1));

        if(overlap)
            return Rcpp::wrap(true);
    }
```

```
41      return Rcpp::wrap(false);
   }
```

Listing 5.13 Function is_overlap from the **wordcloud** package

The package has no other external dependencies and requires only the files created by Rcpp.package.skeleton() as discussed above.

5.4 Further Examples

There are now over 100 packages on the CRAN sites which use **Rcpp**, and which therefore provided working examples which can be studied—just like the **wordcloud** package in the previous section.

Among the CRAN packages using **Rcpp** are

- **RcppArmadillo** (François et al. 2012);
- **RcppEigen** (Bates et al. 2012);
- **RcppBDT** (Eddelbuettel and François 2012b); and,
- **RcppGSL** (François and Eddelbuettel 2010)

all of which not only follow the guidelines described in this chapter but are also discussed in the remainder of the book.

These packages, as well as other packages on CRAN, can serve as examples on how to get data to and from C++ routines and can be considered templates for how to use **Rcpp**. A complete list of packages using **Rcpp** can always found at the CRAN page of the package.

Chapter 6
Extending Rcpp

Abstract This chapter provides an overview of the steps programmers should follow to extend **Rcpp** for use with their own classes and class libraries. The packages **RcppArmadillo**, **RcppEigen**, and **RcppGSL** provide working examples of how to extend **Rcpp** to work with, respectively, the **Armadillo** and **Eigen** C++ class libraries as well as the GNU Scientific Library.

The chapter ends with an illustration of how the **RcppBDT** package connects the date types of R with those of the **Boost Date_Time** library by extending **Rcpp**.

6.1 Introduction

As discussed in the preceding chapters, **Rcpp** facilitates data interchange between R and C++ through the templated function Rcpp::as<>() which convert objects from R to C++, and the function Rcpp::wrap() which converts from C++ to R. In doing so, these function transform data from the representation in the so-called S Expression Pointers (the type SEXP of the R API) to a corresponding (templated) C++ type, and vice versa. The corresponding function declarations for Rcpp::as() and Rcpp::wrap() are as follows:

```
// conversion from R to C++
template <typename T> T as(SEXP m_sexp);

// conversion from C++ to R
template <typename T> SEXP wrap(const T& object);
```

Listing 6.1 as and wrap declarations

These converters are often used implicitly, as in the following example:

```
code <- '
    // we get a list from R
    List input(inputS) ;
```

```
5   // pull std::vector<double> from R list
    // achieved through an implicit call to Rcpp::as
7   std::vector<double> x = input["x"] ;

9   // return an R list
    //  achieved through implicit call to Rcpp::wrap
11  return List::create( _["front"] = x.front(),
                        _["back"]  = x.back());
13  '

15  fx <- cxxfunction(signature(inputS = "list"),
                      body=code, plugin = "Rcpp")
17  input <- list( x = seq(1, 10, by = 0.5) )
    fx( input )
```

Listing 6.2 Implicit use of as and wrap

In this example, a list object (containing a vector x defined as a sequence from 1 to 10) is created in R and passed to the C++ code where a Rcpp::List object is instantiated. A list element named x is then extracted by name and assigned to a C++ vector object. For the return, we also create a list with two named components for the first and last element, respectively, which are named "front" and "back" just like the STL-style accessor function used to extract the corresponding elements.

This example shows how Rcpp::as and Rcpp::wrap can be used to convert standard R and C++ types. These two converter functions have been designed to be extensible to user-defined types and third-party types. The next sections discuss how to apply Rcpp::wrap and Rcpp::as to user-supplied types.

6.2 Extending Rcpp::wrap

The Rcpp::wrap converter is extensible in essentially two ways: a more intrusive approach (which modifies the header files defining the class to be made known to wrap) and via two nonintrusive variants that do not require changes to the class being wrapped. We discuss all three approaches below.

6.2.1 Intrusive Extension

When extending **Rcpp** with your own data type, the recommended way is to implement a conversion to SEXP. This lets Rcpp::wrap know about the new data type. The template meta-programming (or TMP) dispatch is able to recognize that a type is convertible to a SEXP and Rcpp::wrap will then use that conversion.

6.2 Extending Rcpp::wrap

The caveat is that the type must be declared before the main header file Rcpp.h is included.

```
#include <RcppCommon.h>

class Foo {
    public:
        Foo();

        // this operator enables implicit Rcpp::wrap
        operator SEXP();
}

#include <Rcpp.h>
```

Listing 6.3 Intrusive extension for wrap

This is called *intrusive* because the conversion to the SEXP operator has to be declared within the class that we want to use with **Rcpp**. This means we have to add the header RcppCommon.h before the class declaration: this makes SEXP known in the context of our class Foo. By adding the header Rcpp.h later, we ensure that **Rcpp** knows about the conversion from SEXP to foo. And, of course, the actual code for the operator SEXP() has to be supplied as well in a corresponding source file.

6.2.2 Nonintrusive Extension

It is often desirable to offer automatic conversion to third-party types over which the developer has no control. Lack of control, or access to source code, or maybe even design and policy reasons not to alter an existing code base or library may all preclude us from including a conversion to SEXP operator in the class definition as in the previous section.

So to provide automatic conversion from C++ to R, one must declare a specialization of the Rcpp::wrap template between the includes of RcppCommon.h and Rcpp.h.

```
#include <RcppCommon.h>

// third party library that declares class Bar
#include <foobar.h>

// declaring the specialization
namespace Rcpp {
    template <> SEXP wrap( const Bar& );
}

// this must appear after the specialization, else
// the specialization will not be seen by Rcpp types
#include <Rcpp.h>
```

Listing 6.4 Nonintrusive extension for wrap

It should be noted that only the declaration is required. The implementation can appear after the Rcpp.h file is included and can therefore take full advantage of the **Rcpp** type system.

6.2.3 Templates and Partial Specialization

It is also perfectly valid to declare a partial specialization for the Rcpp::wrap template using the templated typename T and a templated use of our class. The compiler will identify the appropriate overload:

```
#include <RcppCommon.h>

// third party library declarings template class Bling<T>
#include <foobar.h>

// declaring the partial specialization
namespace Rcpp {
    namespace traits {
        template <typename T> SEXP wrap( const Bling<T>& );
    }
}

// this must appear after the specialization, else
// the specialization will not be seen by Rcpp types
#include <Rcpp.h>
```

Listing 6.5 Partial specialization for wrap

6.3 Extending Rcpp::as

Conversion from R to C++ using as<>() is also possible in both intrusive and nonintrusive ways.

6.3.1 Intrusive Extension

As part of its template meta-programming dispatch logic, Rcpp::as will attempt to use the constructor of the target class taking a SEXP.

```
#include <RcppCommon.h>

class Foo{
    public:
        Foo() ;
```

6.3 Extending Rcpp::as

```
        // this constructor enables implicit Rcpp::as
        Foo(SEXP) ;
}

#include <Rcpp.h>
```

Listing 6.6 Intrusive extension for `as`

Taking this intrusive route, this constructor can then be implemented in the sources defining class `Foo`.

6.3.2 Nonintrusive Extension

It is also possible to fully specialize `Rcpp::as` to enable nonintrusive implicit conversion capabilities.

```
#include <RcppCommon.h>

// third party library that declares class Bar
#include <foobar.h>

// declaring the specialization
namespace Rcpp {
    template <> Bar as( SEXP ) throw(not_compatible);
}

// this must appear after the specialization, else
// the specialization will not be seen by Rcpp types
#include <Rcpp.h>
```

Listing 6.7 Nonintrusive extension for `as`

6.3.3 Templates and Partial Specialization

The signature of `Rcpp::as` does not allow partial specialization. So when exposing a templated class to `Rcpp::as`, the programmer must specialize the template class `Rcpp::traits::Exporter`. The TMP dispatch will recognize that a specialization of `Exporter` is available and delegate the conversion to this class. **Rcpp** defines the `Rcpp::traits::Exporter` template class as follows:

```
namespace Rcpp {
    namespace traits {

        template <typename T> class Exporter{
        public:
            Exporter( SEXP x ) : t(x){}
```

```
          inline T get(){ return t; }
      private:
          T t;
      } ;
  }
}
```

Listing 6.8 Partial specialization via `Exporter`

This is the reason why the default behavior of `Rcpp::as` is to invoke the constructor of the type `T` taking a `SEXP`.

Since partial specialization of class templates is allowed, we can expose a set of classes as follows:

```
#include <RcppCommon.h>

// third party library that declares template class Bling<T>
#include <foobar.h>

// declaring the partial specialization
namespace Rcpp {
    namespace traits {
        template <typename T> class Exporter< Bling<T> >;
    }
}

// this must appear after the specialization, else
// the specialization will not be seen by Rcpp types
#include <Rcpp.h>
```

Listing 6.9 Partial specialization of as via `Exporter`

Using this approach, the requirements for the `Exporter< Bling<T> >` class are twofold. It should have

- A constructor taking a `SEXP` type.
- A method called `get` which returns an instance of the `Bling<T>` type.

6.4 Case Study: The RcppBDT Package

The package **RcppBDT** (Eddelbuettel and François 2012b) interfaces some of the **Date_Time** classes of the **Boost** C++ library collection.

To do so, it contains `Rcpp::as()` and `Rcpp::wrap()` implementations to convert from one representation to the other. The case discussed in this section is straightforward and concerns the conversion for actual date types which are represented internally as (unsigned) integers.

It does, however, illustrate the general principle of receiving a `SEXP` type and converting to a type from the to-be-wrapped library via the templated function

6.4 Case Study: The **RcppBDT** Package

as<>() in order to convert to a new type, and conversely returning a SEXP via the wrap() function which converts from a new type.

The following specialization includes the actual code along with the declarations.

```
// define template specializations for as and wrap
namespace Rcpp {
    template <> boost::gregorian::date as( SEXP dtsexp ) {
        Rcpp::Date dt(dtsexp);
        return boost::gregorian::date(dt.getYear(),
                                      dt.getMonth(),
                                      dt.getDay());
    }

    template <> SEXP wrap(const boost::gregorian::date &d) {
        boost::gregorian::date::ymd_type ymd =
            d.year_month_day();       // convert to y/m/d struct
        return Rcpp::wrap(Rcpp::Date( ymd.year,
                                      ymd.month,
                                      ymd.day ));
    }
}
```

Listing 6.10 RcppBDT definitions of as and wrap

In the case of as, the SEXP is first converted to a Rcpp::Date type. Its accessors for day, month, and year are then used to instantiate a Boost Gregorian date type. The wrap function here simply does the inverse and deploys the year, month, and date accessors of such a Boost Gregorian date type to access one of the constructor of the Rcpp::Date class.

These two converter functions are then used to pass values between the R representation and the representation used by **Boost Date_Time**.

As an example, consider the implementation of the function

```
Rcpp::Date Date_firstDayOfWeekAfter(boost::gregorian::date *d,
    int weekday, SEXP date) {
    boost::gregorian::first_day_of_the_week_after fdaf(weekday);
    boost::gregorian::date dt =
        Rcpp::as<boost::gregorian::date>(date);
    return Rcpp::wrap(fdaf.get_date(dt));
}
```

Listing 6.11 RcppBDT use of as and wrap

As detailed in Chap. 7, the first argument of function using a so-called "Rcpp Module" has to be a pointer to the object being wrapped, here the class date in the namespace boost::gregorian. The next two arguments are the requested day of the week (encoded as an integer, or using one of constants Mon, Tue, ..., Sun defined in the package) and a date for which the next matching weekday has to be determined.

The function first instantiates a first_day_of_the_week_after object utilizing the **Boost** functionality. Next, the Rcpp::as() converter is used to pass the date obtained from R as a SEXP variable into a date variable, here dt, for

Boost. Finally, Rcpp::wrap() is used to convert the result obtained from computing the "first day of the week after" functionality on the given date dt, returned also as a date object—which Rcpp::wrap converts into a SEXP type from which an Rcpp::Date can be derived implicitly.

This functionality can be illustrated with a simple usage example in which we compute the date of the first Monday after New Year 2020:

```
R> getFirstDayOfWeekAfter(Mon, as.Date("2020-01-01"))
[1] "2020-01-06"
R>
```

Listing 6.12 RcppBDT example for `getFirstDayOfWeekAfter`

Working with *Rcpp modules* is detailed in the next chapter.

6.5 Further Examples

The packages **RcppArmadillo** (François et al. 2012), **RcppEigen** (Bates et al. 2012), and **RcppGSL** (François and Eddelbuettel 2010) provide concrete examples in the context of vector and matrix classes.

Chapter 7
Modules

Abstract This chapter discusses *Rcpp modules* which allow programmers to expose C++ functions and classes to R with relative ease. *Rcpp modules* are inspired from the **Boost.Python** C++ library which provides similar features for integrating Python and C++. Furthermore, *Rcpp modules* also offer the ability to extend C++ classes exposed to R directly from the R side. This chapter discusses modules in detail and ends on an applied case study featuring the **RcppCNPy** package.

7.1 Motivation

Exposing C++ functionality to R is greatly facilitated by the **Rcpp** package and its underlying C++ library. **Rcpp** smoothes many of the rough edges in any R and C++ integration by adding a consistent set of C++ classes to the traditional R Application Programming Interface (API) described in "*Writing R Extensions*" (R Development Core Team 2012d). The **Rcpp**-based approach was the focus of the earlier chapters.

These **Rcpp** facilities offer a lot of assistance to the programmer wishing to interface R and C++. At the same time, they are limited as they operate on a function-by-function basis. The programmer has to implement a `.Call()` compatible function (to conform to the R API) using classes of the **Rcpp** API as we briefly review in the next section.

7.1.1 Exposing Functions Using Rcpp

Exposing existing C++ functions to R through **Rcpp** usually involves several steps. One approach is to write an additional wrapper function that is responsible for converting input objects to the appropriate types, calling the actual worker function and converting the results back to the only suitable type that can be returned to R via the `.Call()` interface: the `SEXP`.

As a concrete example, consider the `norm` function below:

```
double norm( double x, double y ) {
    return sqrt( x*x + y*y );
}
```

Listing 7.1 A simple `norm` function in C++

This simple function does not meet the requirements imposed by the `.Call` interface, so it cannot be called directly by R. Exposing the function involves writing a simple wrapper function that does match the `.Call` interface. **Rcpp** makes this easy.

```
using namespace Rcpp;
RcppExport SEXP norm_wrapper(SEXP x_, SEXP y_) {
    // step 0: convert input to C++ types
    double x = as<double>(x_), y = as<double>(y_);

    // step 1: call the underlying C++ function
    double res = norm( x, y );

    // step 2: return the result as a SEXP
    return wrap( res );
}
```

Listing 7.2 Calling the `norm` function

We use the (templated) **Rcpp** converter `as()` which can transform from a `SEXP` to a number of different C++ and **Rcpp** types; here we used it to assign two scalar `double` types. The **Rcpp** function `wrap()` offers the opposite functionality and converts many known types to a `SEXP`; here we use it to return the `double` scalar result.

This process is simple enough and is widely used by a number of CRAN packages. However, it requires direct involvement from the programmer, which becomes laborious when many functions are involved. *Rcpp modules* provides a much more elegant and unintrusive way to expose C++ functions such as the `norm` function shown above to R.

7.1.2 Exposing Classes Using Rcpp

Exposing C++ classes or structs is even more of a challenge because it requires writing glue code for each member function that is to be exposed.

Consider the simple `Uniform` class below:

```
class Uniform {
public:
    Uniform(double min_, double max_) :
```

7.1 Motivation

```
            min(min_), max(max_) {}

    NumericVector draw(int n) {
        RNGScope scope;
        return runif( n, min, max );
    }

private:
    double min, max;
};
```

Listing 7.3 A simple class `Uniform`

This class enables us to draw a number of uniformly distributed random numbers, and it uses two internal state variables to store the lower and upper bound of the range from which the draws are taken.

To use this class from R, we at least need to expose the constructor and the `draw` method. External pointers (R Development Core Team 2012d) are a suitable mechanism for this, and we can use the `Rcpp::XPtr` template to expose the class with these two functions:

```
using namespace Rcpp;

/// create an external pointer to a Uniform object
RcppExport SEXP Uniform__new(SEXP min_, SEXP max_) {
    // convert inputs to appropriate C++ types
    double min = as<double>(min_), max = as<double>(max_);

    // create a pointer to an Uniform object and wrap it
    // as an external pointer
    Rcpp::XPtr<Uniform> ptr( new Uniform( min, max ), true );

    // return the external pointer to the R side
    return ptr;
}

/// invoke the draw method
RcppExport SEXP Uniform__draw( SEXP xp, SEXP n_ ) {
    // grab the object as a XPtr (smart pointer) to Uniform
    Rcpp::XPtr<Uniform> ptr(xp);

    // convert the parameter to int
    int n = as<int>(n_);

    // invoke the function
    NumericVector res = ptr->draw( n );

    // return the result to R
    return res;
}
```

Listing 7.4 Exposing two member functions for `Uniform` class

However, it is generally considered a bad idea to expose external pointers "as is." Rather, we prefer to have them wrapped as a slot of a corresponding S4 class.

```
R> setClass("Uniform",
+           representation( pointer = "externalptr" ) )
[1] "Uniform"
R> # helper
R> Uniform_method <- function(name) {
+     paste( "Uniform", name, sep = "__" )
+ }
R> # syntactic sugar to allow object$method( ... )
R> setMethod( "$", "Uniform", function(x, name ) {
+     function(...) .Call(Uniform_method(name),
+                         x@pointer, ... )
+ } )
R> # syntactic sugar to allow new( "Uniform", ... )
R> setMethod("initialize",
+           "Uniform", function(.Object, ...) {
+     .Object@pointer <-
+         .Call(Uniform_method("new"), ... )
+     .Object
+ } )
[1] "initialize"
R> u <- new( "Uniform", 0, 10 )
R> u$draw( 10L )
 [1] 4.325224 0.269805 9.990058 7.137135 6.335477
 [5] 6.833734 1.385790 8.850125 1.243403 4.070396
```

Listing 7.5 Using the Uniform class from R

7.2 Rcpp Modules

The design of Rcpp modules has been influenced by **Python modules** which are generated by the **Boost.Python** library (Abrahams and Grosse-Kunstleve 2003). Rcpp modules provide a convenient and easy-to-use way to expose C++ functions and classes to R, grouped together in a single entity.

An Rcpp module is created in a cpp file using the RCPP_MODULE macro, which then provides declarative code of what the module exposes to R.

7.2.1 Exposing C++ Functions Using Rcpp Modules

Consider the norm function from the previous section. We can expose it to R using a single line of code inside the RCPP_MODULE macro:

```
using namespace Rcpp;

double norm( double x, double y ) {
```

7.2 Rcpp Modules

```
4       return sqrt( x*x + y*y );
    }

6   RCPP_MODULE(mod) {
8       function( "norm", &norm );
    }
```

Listing 7.6 Exposing the norm function via modules

The code creates an Rcpp module called mod that exposes the norm function. **Rcpp** automatically deduces the conversions that are needed for input and output. This alleviates the need for a wrapper function using either **Rcpp** or the R API.

On the R side, the module is retrieved by using the Module() function from **Rcpp**:

```
1   R> require( Rcpp )
    R> mod <- Module( "mod" )
3   R> mod$norm( 3, 4 )
```

Listing 7.7 Using norm function exposed via modules

A module can contain any number of calls to function to register many internal functions to R. For example, consider these six functions covering a range of input and return arguments:

```
1   std::string hello() { return "hello"; }

3   int bar( int x) { return x*2; }

5   double foo( int x, double y) { return x * y; }

7   void bla( ) { Rprintf( "hello\\n" ); }

9   void bla1( int x) {
        Rprintf( "hello (x = %d)\\n", x );
11  }

13  void bla2( int x, double y) {
        Rprintf( "hello (x = %d, y = %5.2f)\\n", x, y );
15  }
```

Listing 7.8 A module example with six functions

They can all be exposed to R with the following minimal code:

```
1   RCPP_MODULE(yada) {
        using namespace Rcpp;
3
        function( "hello"  , &hello  );
5       function( "bar"    , &bar    );
        function( "foo"    , &foo    );
7       function( "bla"    , &bla    );
        function( "bla1"   , &bla1   );
```

```
      function( "bla2"  , &bla2  );
9
}
```

Listing 7.9 Modules example interface

We can now use them from R as follows:

```
R> require( Rcpp )
2 R> yada <- Module( "yada" )
R> yada$bar( 2L )
4 R> yada$foo( 2L, 10.0 )
R> yada$hello()
6 R> yada$bla()
R> yada$bla1( 2L)
8 R> yada$bla2( 2L, 5.0 )
```

Listing 7.10 Modules example use from R

The requirements for a function to be exposed to R via Rcpp modules are as follows:

- The function has to take between 0 and 65 parameters.
- Each input parameter must be manageable by the templated Rcpp::as conversion function.
- The return type of the function must be either void or any type that can be managed by the Rcpp::wrap template conversion function.
- The function name itself has to be unique in the module. In other words, no two functions with the same name but different signatures are allowed. While C++ allows overloading functions, Rcpp modules relies on named identifiers for the lookup and cannot allow two identical identifiers.

7.2.1.1 Documentation for Exposed Functions Using Rcpp Modules

In addition to the name of the function and the function pointer, it is possible to pass a short description of the function as the third parameter of function.

```
using namespace Rcpp;
2
double norm( double x, double y ) {
4    return sqrt( x*x + y*y );
}
6
RCPP_MODULE(mod) {
8    function("norm", &norm,
             "Provides a simple vector norm" );
10 }
```

Listing 7.11 Modules example with function documentation

7.2 Rcpp Modules   89

The description is used when displaying the function to the R prompt:

```
R> mod <- Module( "mod", getDynLib( fx ) )
R> show( mod$norm )
internal C++ function <0x2477630>
    docstring  : Provides a simple vector norm
    signature  : double norm(double, double)
```

Listing 7.12 Output for modules example with function documentation

7.2.1.2 Formal Arguments Specification

Using function, we can specify the formal arguments of the R function that encapsulates the C++ function by passing a Rcpp::List after the function pointer and before the (also optional) documentation entry:

```
using namespace Rcpp;

double norm( double x, double y ) {
    return sqrt( x*x + y*y );
}

RCPP_MODULE(mod_formals) {
    function("norm",
             &norm,
             List::create(_["x"] = 0.0, _["y"] = 0.0),
             "Provides a simple vector norm");
}
```

Listing 7.13 Modules example with documentation and formal arguments

A simple usage example is provided below:

```
R> norm <- mod$norm
R> norm()
[1] 0
R> norm( y = 2 )
[1] 2
R> norm( x = 2, y = 3 )
[1] 3.605551
R> args( norm )
function (x = 0, y = 0)
NULL
```

Listing 7.14 Output for modules example with documentation and formal arguments

To set formal arguments without default values, simply omit the right-hand side.

```
using namespace Rcpp;

double norm( double x, double y ) {
    return sqrt( x*x + y*y );
}
```

```
RCPP_MODULE(mod_formals2) {
    function( "norm", &norm,
              List::create( _["x"], _["y"] = 0.0),
              "Provides a simple vector norm");
}
```

Listing 7.15 Modules example with documentation and formal arguments without defaults

This can be used as follows:

```
R> norm <- mod$norm
R> args( norm )
function (x, y = 0)
NULL
```

Listing 7.16 Usage of modules example with documentation and formal arguments

The ellipsis (...) can be used to denote that additional arguments are optional; it does not take a default value.

```
using namespace Rcpp;

double norm( double x, double y ) {
    return sqrt( x*x + y*y );
}

RCPP_MODULE(mod_formals3) {
    function( "norm", &norm,
              List::create( _["x"], _["..."] ),
              "documentation for norm");
}
```

Listing 7.17 Modules example with ellipis argument

This now shows the ellipsis in the documentation output.

```
R> norm <- mod$norm
R> args( norm )
function (x, ...)
NULL
```

Listing 7.18 Output of modules example with ellipis argument

7.2.2 Exposing C++ Classes Using Rcpp Modules

Rcpp modules also provide a mechanism for exposing C++ classes, based on the Reference Classes which were first introduced in R release 2.12.0.

7.2.2.1 Initial Example

A class is exposed using the `class_` keyword (and the trailing underscore is required as we cannot use the C++ language keyword `class`). The `Uniform` class may be exposed to R as follows:

```cpp
using namespace Rcpp;
class Uniform {
public:
    Uniform(double min_, double max_) :
        min(min_), max(max_) {}

    NumericVector draw(int n) const {
        RNGScope scope;
        return runif( n, min, max );
    }

    double min, max;
};

double uniformRange( Uniform* w) {
    return w->max - w->min;
}

RCPP_MODULE(unif_module) {

    class_<Uniform>( "Uniform" )

    .constructor<double,double>()

    .field( "min", &Uniform::min )
    .field( "max", &Uniform::max )

    .method( "draw", &World::draw )
    .method( "range", &uniformRange )
    ;

}
```

Listing 7.19 Exposing `Uniform` class using modules

A short example follows and shows how to use this class:

```
R> Uniform <- unif_module$Uniform
R> u <- new( Uniform, 0, 10 )
R> u$draw( 10L )
 [1] 3.7950482 6.9525034 0.5783621 5.7234278 0.6869314
 [6] 5.6403064 2.3408875 6.5695670 1.8821565 8.8553301
R> u$range()
[1] 10
R> u$max <- 1
R> u$range()
[1] 1
R> u$draw( 10 )
```

```
12  [1] 0.1987632 0.7598329 0.7276362 0.3101182 0.2300929
    [6] 0.7121408 0.1005060 0.4007011 0.1643178 0.2252207
```

Listing 7.20 Using Uniform class via modules

Here, `class_` is templated by the C++ class or struct that is to be exposed to R. The parameter of the `class_<Uniform>` constructor is the name we will use on the R side. It often makes sense to use the same name as the class name. While this is not enforced, it might be useful when exposing a class generated from a template.

Then a single constructor, two fields and two methods are exposed to complete the simple example. Of the two methods, one accesses a class member function (`draw`), whereas the other uses a free function (`uniformRange`).

7.2.2.2 Exposing Constructors Using Rcpp Modules

Public constructors that take from zero and seven parameters can be exposed to the R level using the `.constuctor` template method of `.class_`.

Optionally, `.constructor` can take a description as the first argument.

```
1  .constructor<double, double>(
       "sets the min and max value of the distribution")
```

Listing 7.21 Constructor with a description

Also, the second argument can be a function pointer (called validator) matching the following type :

```
typedef bool (*ValidConstructor)(SEXP*, int);
```

Listing 7.22 Constructor with a validator function pointer

The validator can be used to implement dispatch to the appropriate constructor, when multiple constructors taking the same number of arguments are exposed. The default validator always accepts the constructor as valid if it is passed the appropriate number of arguments. For example, with the call above, the default validator accepts any call from R with two `double` arguments (or arguments that can be cast to `double`).

7.2.2.3 Exposing Fields and Properties

`class_` has three ways to expose fields and properties, as illustrated in the example below:

```
1  using namespace Rcpp;
   class Foo {
3      public:
           Foo( double x_, double y_, double z_ ):
5              x(x_), y(y_), z(z_) {}
```

7.2 Rcpp Modules

```
            double x;
            double y;

            double get_z() { return z; }
            void set_z( double z_ ) { z = z_; }

        private:
            double z;
    };

    RCPP_MODULE(mod_foo) {
        class_<Foo>( "Foo" )

        .constructor<double,double,double>()

        .field( "x", &Foo::x )
        .field_readonly( "y", &Foo::y )

        .property( "z", &Foo::get_z, &Foo::set_z )
        ;
    }
```

Listing 7.23 Exposing fields and properties for modules

The .field method exposes a public field with read/write access from R; field also accepts an extra parameter to give a short description of the field:

```
.field( "x", &Foo::x, "documentation for x" )
```

Listing 7.24 Field with documentation

The .field_readonly method exposes a public field with read-only access from R. It also accepts the description of the field.

```
.field_readonly( "y", &Foo::y, "documentation for y" )
```

Listing 7.25 Readonly-field with documentation

The .property method allows indirect access to fields through a getter and a setter function. The setter is optional, and the property is considered read-only if the setter is not supplied. As before, an optional documentation string can also be supplied to describe the property:

```
// with getter and setter
.property("z", &Foo::get_z, &Foo::set_z,
          "Documentation for z" )

// with only a getter
.property( "z", &Foo::get_z, "Documentation for z" )
```

Listing 7.26 Property with getter and setter, or getter-only

The type of the field (T) is deduced from the return type of the getter, and if a setter is given, its unique parameter should be of the same type.

Getters can be member functions taking no parameters and returning a T (e.g., get_z above), or a free function taking a pointer to the exposed class and returning a T, for example:

```
double z_get( Foo* foo ) { return foo->get_z(); }
```

Listing 7.27 Example of using a getter

Setters can be either a member function taking a T and returning void, such as set_z above, or a free function taking a pointer to the target class and a T:

```
void z_set( Foo* foo, double z ) { foo->set_z(z); }
```

Listing 7.28 Example of using a setter

Using properties gives more flexibility in case field access has to be tracked or has an impact on other fields. For example, this class keeps track of how many times the x field is read and written.

```
class Bar {
    public:

        Bar(double x_) : x(x_), nread(0), nwrite(0) {}

        double get_x( ) {
            nread++;
            return x;
        }

        void set_x( double x_ ) {
            nwrite++;
            x = x_;
        }

        IntegerVector stats() const {
            return IntegerVector::create(
                _["read"] = nread,
                _["write"] = nwrite
            );
        }

    private:
        double x;
        int nread, nwrite;
};

RCPP_MODULE(mod_bar) {
    class_<Bar>( "Bar" )

        .constructor<double>()
```

7.2 Rcpp Modules

```
33            .property( "x", &Bar::get_x, &Bar::set_x )
              .method( "stats", &Bar::stats )
35        ;
    }
```

Listing 7.29 Example code for properties

Here is a simple usage example:

```
  R> Bar <- mod_bar$Bar
2 R> b <- new( Bar, 10 )
  R> b$x + b$x
4 [1] 20
  R> b$stats()
6   read write
         2    0
8 R> b$x <- 10
  R> b$stats()
10  read write
         2    1
```

Listing 7.30 Example using properties

7.2.2.4 Exposing Methods Using Rcpp Modules

`class_` has several overloaded and templated `.method` functions allowing the programmer to expose a method associated with the class.

A legitimate method to be exposed by `.method` can be:

- A public member function of the class, either const or non-const, that returns void or any type that can be handled by `Rcpp::wrap`, and that takes between 0 and 65 parameters whose types can be handled by `Rcpp::as.`; or
- A free function that takes a pointer to the target class as its first parameter, followed by 0 or more (up to 65) parameters that can be handled by `Rcpp::as` and returning a type that can be handled by `Rcpp::wrap` or void.

Documenting Methods

`.method` can also include a short documentation of the method, after the method (or free function) pointer.

```
1 .method( "stats", &Bar::stats,
           "vector indicating the number of times"
3          "x has been read and written" )
```

Listing 7.31 Example documenting a method

Note that the documentation string is really only one argument as there is no comma separating the two pieces.

Const and Non-const Member Functions

`method` is able to expose both `const` and `non-const` member functions of a class. There are, however, situations where a class defines two versions of the same method, differing only in their signature by the `const`-ness. This is, for example, the case of the member functions `back` of the `std::vector` template from the STL.

```
reference back ( );
const_reference back ( ) const;
```

Listing 7.32 Const and non-const member functions

To resolve the ambiguity, it is possible to use either the `const_method` or the `nonconst_method` instead of `method` in order to restrict the candidate methods.

Special Methods

Rcpp considers the methods `[[` and `[[<-` special and promotes them to indexing methods on the R side.

7.2.2.5 Object Finalizers

The `.finalizer` member function of `class_` can be used to register a finalizer. A finalizer is a free function that takes a pointer to the target class and returns `void`. The finalizer is called before the destructor and so operates on a valid object of the target class. It can be used to perform suitable operations such as releasing resources, or summarizing and logging behavior.

The finalizer is called automatically when the R object that encapsulates the C++ object is garbage-collected.

7.2.2.6 S4 Dispatch

When a C++ class is exposed by the `class_` template, a new S4 class is registered as well. The name of the S4 class is obfuscated in order to avoid name clashes (i.e., two modules exposing the same class).

This allows for the implementation of R-level (S4) dispatch. For example, one might implement the `show` method for C++ `World` objects:

```
setMethod("show", yada$World,
          function(object) {
              msg <- paste("World object with message :",
                           object$greet() )
              writeLines( msg )
          } )
```

Listing 7.33 Example of S4 dispatch

7.2 Rcpp Modules

7.2.2.7 Full Example

The following example illustrates how to use Rcpp modules to expose the class `std::vector<double>` from the STL.

```cpp
// convenience typedef
typedef std::vector<double> vec;

// helpers
void vec_assign( vec* obj, Rcpp::NumericVector data ) {
    obj->assign( data.begin(), data.end() );
}

void vec_insert(vec* obj, int position,
                Rcpp::NumericVector data) {
    vec::iterator it = obj->begin() + position;
    obj->insert( it, data.begin(), data.end() );
}

Rcpp::NumericVector vec_asR( vec* obj ) {
    return Rcpp::wrap( *obj );
}

void vec_set( vec* obj, int i, double value ) {
    obj->at( i ) = value;
}

RCPP_MODULE(mod_vec) {
    using namespace Rcpp;

    // we expose the class std::vector<double>
    // as "vec" on the R side
    class_<vec>( "vec")

    // exposing constructors
    .constructor()
    .constructor<int>()

    // exposing member functions
    .method( "size", &vec::size)
    .method( "max_size", &vec::max_size)
    .method( "resize", &vec::resize)
    .method( "capacity", &vec::capacity)
    .method( "empty", &vec::empty)
    .method( "reserve", &vec::reserve)
    .method( "push_back", &vec::push_back )
    .method( "pop_back", &vec::pop_back )
    .method( "clear", &vec::clear )

    // specifically exposing const member functions
    .const_method( "back", &vec::back )
    .const_method( "front", &vec::front )
    .const_method( "at", &vec::at )
```

```
50      // exposing free functions taking a
        // std::vector<double>* as their first argument
52      .method( "assign", &vec_assign )
        .method( "insert", &vec_insert )
54      .method( "as.vector", &vec_asR )

56      // special methods for indexing
        .const_method( "[[", &vec::at )
58      .method( "[[<-", &vec_set )

60      ;
    }
```

Listing 7.34 Complete example of exposing `std::vector<double>`

The R usage is as follows:

```
1  R> vec <- mod_vec$vec
   R> v <- new( vec )
3  R> v$reserve( 50L )
   R> v$assign( 1:10 )
5  R> v$push_back( 10 )
   R> v$size()
7  R> v$capacity()
   R> v[[ 0L ]]
9  R> v$as.vector()
```

Listing 7.35 R use of `std::vector<double>` modules example

7.3 Using Modules in Other Packages

7.3.1 Namespace Import/Export

7.3.1.1 Import All Functions and Classes

When using **Rcpp** modules in a package, the client package needs to import **Rcpp**'s namespace. This is achieved by adding the following line to the NAMESPACE file.

```
1  import( Rcpp )
```

Listing 7.36 R NAMESPACE import of **Rcpp** for modules

Loading the module must happen after the dynamic library of the package is loaded. There are two approaches. The older one uses the `.onLoad()` hook.

```
1  # grab the namespace
   NAMESPACE <- environment()
3
   .onLoad <- function(libname, pkgname) {
5      ## load the module and store it in our namespace
```

7.3 Using Modules in Other Packages

```
    yada <- Module( "yada" )
    populate( yada, NAMESPACE )
}
```

Listing 7.37 R .onLoad() code for module

The call to `populate` installs all functions and classes from the module into the namespace of package.

7.3.1.2 Import All Modules in a Package

There is also a convenience function `loadRcppModules()` that loops over all modules declared in the DESCRIPTION file. The `loadRcppModules()` function has a single argument `direct` with a default value of TRUE implying that all (exported) identifiers in the module are directly populated into the namespace of the module. Otherwise, only the module is exposed and its functions need to be addressed indirectly (as, for example, via `v$size()`).

The `loadRcppModules()` function has to be called from the `.onLoad()` function as well.

7.3.1.3 Using loadModule

A separate function `loadModule()` has been available since release 0.9.11 of **Rcpp**. It can be used in any .R function in a package (and not just in .onLoad()). Its first argument is the module name. The second argument can be used to detail which parts of a module should be loaded; the special value TRUE signals that all objects with valid names are exported.

As an example, **RcppBDT** uses `loadModule("bdtDtMod", TRUE)` to load all components of the bdtDtMod module. Similarly, the **RcppCNPy** package uses `loadModule("cnpy", TRUE)` to load its sole module cnpy.

7.3.2 Support for Modules in Skeleton Generator

The `Rcpp.package.skeleton()` function has been extended to facilitate the use of **Rcpp** modules. When the module argument is set to TRUE, the skeleton generator installs code that uses a simple module.

```
R> Rcpp.package.skeleton( "testmod", module = TRUE )
```

Listing 7.38 Package skeleton support for modules

This provides the easiest way to create a new package containing Rcpp modules code.

7.3.3 Module Documentation

Rcpp defines a `prompt()` method for the `Module` class, allowing generation of a skeleton of an Rd file containing some information about the module.

```
R> yada <- Module( "yada" )
R> prompt( yada, "yada-module.Rd" )
```

Listing 7.39 Use of `prompt` for documentation skeleton

7.4 Case Study: The RcppCNPy Package

Modules are a very powerful tool. They are suited to exposing code from existing class libraries as the **RcppBDT** package discussed in the previous section illustrates. The Rcpp attributes system uses modules to easily connect the wrappers it generates for the user-supplied code.

Modules can be used to provide simple wrappers to external libraries. A simple example is provided by the **RcppCNPy** package (Eddelbuettel 2012a). It uses a small stand-alone library provided in a single header and source files in order to access NumPy files used by this popular **Python** extension.

In the package, two functions `npyLoad()` and `npySave()` are defined. They simply transfer data between a given file name and R by relying on the external library provided by the source files `cnpy.cpp` and `cnpy.h`.

Listing 7.40 shows simplified versions of these two functions. We have omitted several aspects to keep the exposition shorter: the special case of transposing, the additional layer of transparently dealing with `gzip`-compressed files, the support for `long long` types, as well as the support for different underlying data types concentratic on just `numeric`.

```
Rcpp::RObject npyLoad(const std::string & filename,
                      const std::string & type) {
    cnpy::NpyArray arr;
    arr = cnpy::npy_load(filename);

    std::vector<unsigned int> shape = arr.shape;
    SEXP ret = R_NilValue;
    if (shape.size() == 1) {
        if (type == "numeric") {
            double *p = reinterpret_cast<double*>(arr.data);
            ret = Rcpp::NumericVector(p, p + shape[0]);
        } else {
            arr.destruct();
            Rf_error("Unsupported type in npyLoad");
        }
    } else if (shape.size() == 2) {
        if (type == "numeric") {
            ret = Rcpp::NumericMatrix(shape[0], shape[1],
```

7.4 Case Study: The RcppCNPy Package 101

```
                        reinterpret_cast<double*>(arr.data)));
        } else {
            arr.destruct();
            Rf_error("Unsupported type in npyLoad");
        }
    } else {
        arr.destruct();
        Rf_error("Unsupported dimension in npyLoad");
    }
    arr.destruct();
    return ret;
}

void npySave(std::string filename, Rcpp::RObject x,
             std::string mode) {
    if (::Rf_isMatrix(x)) {
        if (::Rf_isNumeric(x)) {
            Rcpp::NumericMatrix mat =
                transpose(Rcpp::NumericMatrix(x));
            std::vector<unsigned int> shape =
                Rcpp::as<std::vector<unsigned int> >(
                    Rcpp::IntegerVector::create(mat.ncol(),
                                                mat.nrow()));

            cnpy::npy_save(filename, mat.begin(),
                           &(shape[0]), 2, mode);
        } else {
            Rf_error("Unsupported matrix type\n");
        }
    } else if (::Rf_isVector(x)) {
        if (::Rf_isNumeric(x)) {
            Rcpp::NumericVector vec(x);
            std::vector<unsigned int> shape =
                Rcpp::as<std::vector<unsigned int> >(
                    Rcpp::IntegerVector::create(vec.length()));
            cnpy::npy_save(filename, vec.begin(),
                           &(shape[0]), 1, mode);
        } else {
            Rf_error("Unsupported vector type\n");
        }
    } else {
        Rf_error("Unsupported type\n");
    }
}
```

Listing 7.40 NumPy load and save functions defined in **RcppCNPy**

The following module declaration is all it takes to make the functions known to R along with some standard declaration done by the rest of the package, and provided by helper functions like `Rcpp.package.skeleton()` when the `modules=TRUE` option is selected.

```
RCPP_MODULE(cnpy){

    using namespace Rcpp;

    // name of the identifier at the R level
    function("npyLoad",
             // function pointer to helper function defined above
             &npyLoad,
             // function arguments including default value
             List::create( Named("filename"),
                           Named("type") = "numeric",
                           Named("dotranspose") = true),
             "read an npy file into a numeric vector or matrix");

    // name of the identifier at the R level
    function("npySave",
             // function pointer to helper function defined above
             &npySave,
             // function arguments including default value
             List::create( Named("filename"),
                           Named("object"),
                           Named("mode") = "w"),
             "save an R object to an npy file");

}
```

Listing 7.41 Example of module declaration in **RcppCNPy**

7.5 Further Examples

Several packages on CRAN now use Rcpp modules. As of late 2012, the list comprises the packages **GUTS** (Albert and Vogel 2012), **RSofia** (King and Diaz 2011), **RcppBDT** (Eddelbuettel and François 2012b), **RcppCNPy** (Eddelbuettel 2012a), **cda** (Auguie 2012a), **highlight** (François 2012a), **maxent** (Jurka and Tsuruoka 2012), **parser** (François 2012b), **planar** (Auguie 2012b), and **transmission** (Thomas and Redd 2012).

Chapter 8
Sugar

Abstract This chapter describes *Rcpp sugar* which brings a higher level of abstraction to C++ code written using the **Rcpp** API. *Rcpp sugar* is based on expression templates and provides some "syntactic sugar" facilities directly in **Rcpp**. In this chapter, we will introduce many of the very useful *Rcpp sugar* features. As our focus is firmly on using *Rcpp sugar*, we will do so without venturing too deeply into the template meta programming approach used to implement it. Some technical details are provided at the end, and this section can be skipped by users who are interested primarily in using, rather than extending, *Rcpp sugar*. A brief simulation example using *Rcpp sugar* concludes the chapter.

8.1 Motivation

Rcpp facilitates development of compiled code to extend R via either inline code or an R package by abstracting low-level details of the R API (R Development Core Team 2012d) into a consistent set of C++ classes.

Code written using **Rcpp** classes is easier to read, write, and maintain, without losing performance. Consider the following code example which provides a function `foo` as a C++ extension to R by using the **Rcpp** API:

```
RcppExport SEXP foo( SEXP xs, SEXP ys) {
    Rcpp::NumericVector xv(xs);
    Rcpp::NumericVector yv(ys);
    int n = xv.size();
    Rcpp::NumericVector res( n );
    for (int i=0; i<n; i++) {
        double x = xv[i];
        double y = yv[i];
        if( x < y ) {
            res[i] = x * x;
        } else {
```

```
            res[i] = -( y * y );
        }
    }
    return res;
}
```

Listing 8.1 A simple C++ function operating on vectors

The goal of the function `foo` code is simple. We pass two `numeric` vectors passed from R (as `SEXP` types) and create two **Rcpp** vectors. We then create a third one of the same length as x, fill it, and return it to R (and let us ignore for a moment that the actual transformation from `xv` and `yv` into the results vector `res` is not all that meaningful). This function shows typical low-level C++ code that could be written much more concisely in R, thanks to vectorization as shown in the next example.

```
R> foo <- function(x, y){
+    ifelse( x < y, x*x, -(y*y) )
+ }
```

Listing 8.2 A simple R function operating on vectors

Put succinctly, the motivation of *Rcpp sugar* is to bring a subset of the high-level R syntax to C++. Hence, with *Rcpp sugar*, the C++ version of `foo` now becomes

```
RcppExport SEXP foo( SEXP xs, SEXP ys){
    Rcpp::NumericVector x(xs) ;
    Rcpp::NumericVector y(ys) ;
    return ifelse( x < y, x*x, -(y*y) ) ;
}
```

Listing 8.3 A simple C++ function using sugar operating on vectors

which is only about a third as long as the initial version. More importantly, it permits us to collapse explicit loops (which are common in C++ but we note that, for example, the STL offers alternatives) into vectorized expressions just as it would in R.

Apart from the fact that we need to assign the two objects we obtain from R— which is a simple statement each thanks to the template magic in **Rcpp**, and as previously discussed also a lightweight operation copying only a pointer—and the need for an explicit `return` statement, the code is now identical between highly vectorized R and C++. So *Rcpp sugar* enables us to express a vectorized expression in C++ just as easily as in R.

Rcpp sugar is written using expression templates and lazy evaluation techniques (Abrahams and Gurtovoy 2004; Vandevoorde and Josuttis 2003). This not only allows for a much nicer high-level syntax but also makes it rather efficient as we detail further in Sects. 8.4 and 8.6 below.

8.2 Operators

Rcpp sugar takes advantage of C++ operator overloading. The next few sections discuss several examples.

8.2.1 Binary Arithmetic Operators

Rcpp sugar defines the usual binary arithmetic operators: +, -, *, /.

```
// two numeric vectors of the same size
NumericVector x;
NumericVector y;

// expressions involving two vectors
NumericVector res = x + y;
NumericVector res = x - y;
NumericVector res = x * y;      // NB element-wise multiplication
NumericVector res = x / y;

// one vector, one single value
NumericVector res = x + 2.0;
NumericVector res = 2.0 - x;
NumericVector res = y * 2.0;
NumericVector res = 2.0 / y;

// two expressions
NumericVector res = x * y + y / 2.0;
NumericVector res = x * ( y - 2.0 );
NumericVector res = x / ( y * y );
```

Listing 8.4 Binary arithmetic operators for sugar

The left-hand side (lhs) and the right-hand side (rhs) of each binary arithmetic expression must be of the same type (e.g., they should be both numeric expressions).

The lhs and the rhs can either have the same size or one of them could be a primitive value of the appropriate type, for example adding a NumericVector and a double. This is different from R which uses a recycling rule for its operation. When a shorter vector, say, of length four is added to a longer vector of length eight, the recycling operation can succeed as an integer multiple (here: two) of the shorter vector's length (here: four) is equal to the longer vector's length (here: eight). This behavior is not emulated in *Rcpp sugar* where either the two operants have to be of the same length or one has to be a single primitive C++ type such as double.

8.2.2 Binary Logical Operators

Binary logical operators create a `logical` sugar expression, from either two sugar expressions of the same type or one sugar expression and a primitive value of the associated type.

```
  // two integer vectors of the same size
2 NumericVector x;
  NumericVector y;
4
  // expressions involving two vectors
6 LogicalVector res = x < y;
  LogicalVector res = x > y;
8 LogicalVector res = x <= y;
  LogicalVector res = x >= y;
10 LogicalVector res = x == y;
  LogicalVector res = x != y;
12
  // one vector, one single value
14 LogicalVector res = x < 2;
  LogicalVector res = 2 > x;
16 LogicalVector res = y <= 2;
  LogicalVector res = 2 != y;
18
  // two expressions
20 LogicalVector res = ( x + y ) <  ( x*x );
  LogicalVector res = ( x + y ) >= ( x*x );
22 LogicalVector res = ( x + y ) == ( x*x );
```

Listing 8.5 Binary logical operators for sugar

8.2.3 Unary Operators

The unary `operator-` can be used to negate a (numeric) sugar expression, whereas the unary `operator!` negates a logical sugar expression:

```
  // a numeric vector
2 NumericVector x;

4 // negate x
  NumericVector res = -x;
6
  // use it as part of a numerical expression
8 NumericVector res = -x * ( x + 2.0 );

10 // two integer vectors of the same size
  NumericVector y;
12 NumericVector z;

14 // negate the logical expression "y < z"
  LogicalVector res = ! ( y < z );
```

Listing 8.6 Unary operators for sugar

8.3 Functions

Rcpp sugar defines functions that closely match the behavior of R functions of the same name.

8.3.1 Functions Producing a Single Logical Result

Given a logical sugar expression, the `all` function identifies if all the elements are TRUE. Similarly, the `any` function identifies if any one of the elements of a given logical sugar expression is TRUE.

```
1 IntegerVector x = seq_len( 1000 );
  all( x*x < 3 );
3 any( x*x < 3 );
```

Listing 8.7 Functions returning a single boolean result

Either call to `all` and `any` creates an object of a class that has member functions `is_true`, `is_false`, `is_na` and a conversion to SEXP operator.

One important thing to highlight is that `all` is lazy. Unlike in R, there is no need to fully evaluate the expression. In the example above, the result of `all` is fully resolved after evaluating only the first two indices of the expression x * x < 3. `any` is lazy too, so it will only need to resolve the first element of the example above.

Another important consideration is the conversion to the `bool` type. In order to respect the concept of missing values (NA) in R, expressions generated by `any` or `all` cannot be converted directly to `bool`. Instead one must use `is_true`, `is_false` or `is_na`:

```
1 // wrong: will generate a compile error
  bool res = any( x < y) );
3
  // ok
5 bool res = is_true( any( x < y ) );
  bool res = is_false( any( x < y ) );
7 bool res = is_na( any( x < y ) );
```

Listing 8.8 Using functions returning a single boolean result

8.3.2 Functions Producing Sugar Expressions

8.3.2.1 is_na

Given a sugar expression of any type, `is_na` (just like the other functions in this section) produces a logical sugar expression of the same length. Each element of the result expression evaluates to TRUE if the corresponding input is a missing value, or FALSE otherwise.

```
1  IntegerVector x = IntegerVector::create( 0, 1, NA_INTEGER, 3 );

3  is_na( x );
   all( is_na( x ) );
5  any( ! is_na( x ) );
```

Listing 8.9 Example using is_na sugar function

8.3.2.2 seq_along

Given a sugar expression of any type, seq_along creates an integer sugar expression whose values go from 1 to the size of the input.

```
1  IntegerVector x = IntegerVector::create( 0, 1, NA_INTEGER, 3 );

3  seq_along( x );
   seq_along( x * x * x * x * x * x * x );
```

Listing 8.10 Example using seq_along sugar function

This is a "lazy" (in the R evaluation sense) function, as it only needs to call the size member function of the input expression. In other words, the value of the input expression does need not to be computed. The two examples above give the same result with the same efficiency at run-time. The compile time will be affected by the complexity of the second expression, since the abstract syntax tree is built at compile time.

8.3.2.3 seq_len

seq_len creates an integer sugar expression whose *i*th element expands to i. This makes seq_len particularly useful for functions such as sapply and lapply (which are similar to their R equivalents, and discussed below).

```
   // 1, 2, ..., 10
2  IntegerVector x = seq_len( 10 );

4  lapply( seq_len(10), seq_len );
```

Listing 8.11 Example using seq_len sugar function

8.3.2.4 pmin and pmax

Given two sugar expressions of the same type and size, or one expression and one primitive value of the appropriate type, pmin (pmax) generates a sugar expression

8.3 Functions

of the same type whose *i*th element expands to the lowest (highest) value between the *i*th element of the first expression and the *i*th element of the second expression.

```
  IntegerVector x = seq_len( 10 );
2
  pmin( x, x*x );
4 pmin( x*x, 2 );

6 pmin( x, x*x );
  pmin( x*x, 2 );
```

Listing 8.12 Example using `pmin` and `pmax` sugar function

8.3.2.5 ifelse

Given a logical sugar expression and either

- Two compatible sugar expressions (same type, same size) or
- One sugar expression and one compatible primitive.

`ifelse` expands to a sugar expression whose *i*th element is the *i*th element of the first expression if the *i*th element of the condition expands to TRUE, or the *i*th of the second expression if the *i*th element of the condition expands to FALSE, or the appropriate missing value otherwise.

```
1 IntegerVector x;
  IntegerVector y;
3
  ifelse( x < y, x, (x+y)*y );
5 ifelse( x > y, x, 2 );
```

Listing 8.13 Example using `ifelse` sugar function

8.3.2.6 sapply

`sapply` applies a C++ function to each element of the given expression to create a new expression. The type of the resulting expression is deduced by the compiler from the result type of the function.

The function can be a free C++ function such as the overload generated by the template function below:

```
1 template <typename T>
  T square( const T& x) {
3     return x * x;
  }
5 sapply( seq_len(10), square<int> );
```

Listing 8.14 Example using `sapply` sugar function

Alternatively, the function can be a functor whose type has a nested type called `result_type`. One way to satisfy this requirement is by inheriting from the `std::unary_function` functor:

```
template <typename T>
struct square : std::unary_function<T,T> {
    T operator()(const T& x){
        return x * x;
    }
}
sapply( seq_len(10), square<int>() );
```

Listing 8.15 Example using `std::unary_function` functor with `sapply`

8.3.2.7 lapply

`lapply` is similar to `sapply` except that the result is always a list expression (an expression of type `VECSXP`).

8.3.2.8 mapply

`mapply` is similar to `sapply` and `lapply` but permits multiple vectors as input. This is (at least currently) limited to either two or three vectors.

We can modify the example from Listing 8.14 to illustrate how `mapply` can be used to work on multiple vectors. Here, instead of computing the squared value of each element, we "sweep" a sum of squares calculation across two vectors.

```
template <typename T>
struct sumOfSquares : std::unary_function<T,T> {
    T operator()(const T& x, const T& y){
        return x*x + y*y;
    }
}

NumericVector res;
res = mapply(seq_len(10), seq_len(10),
             sumOfSquares<double>() );
```

Listing 8.16 Example using `std::unary_function` functor with `mapply`

8.3.2.9 sign

Given a numeric or integer expression, `sign` expands to an expression whose values are one of 1, 0, −1, or `NA`, depending on the sign of the input expression.

8.3 Functions

```
  IntegerVector xx;

  sign( xx );
  sign( xx * xx );
```

Listing 8.17 Example using `sign` sugar function

8.3.2.10 diff

The *i*th element of the result of `diff` is the difference between the $(i+1)$th and the *i*th element of the input expression. Supported types are integer and numeric.

```
  IntegerVector xx;

  diff( xx );
```

Listing 8.18 Example using `sign` sugar function

8.3.2.11 setdiff

The `setdiff` function returns the values of the first vector which are not contained in the second vector; this is analogous to the R version.

```
  IntegerVector xx, yy;

  setdiff( xx, yy );
```

Listing 8.19 Example using `setdiff` sugar function

8.3.2.12 union_

The `union_` function returns the union of the two vectors. The function has to be named with the trailing underscore in order not to conflict with the language keyword `union`.

```
  IntegerVector xx, yy;

  union_( xx, yy );
```

Listing 8.20 Example using `union_` sugar function

8.3.2.13 intersect

The `intersect` function returns the intersection of the two vectors.

```
IntegerVector xx, yy;

intersect( xx, yy );
```
Listing 8.21 Example using `intersect` sugar function

8.3.2.14 clamp

The `clamp` function combines the application of both `pmin` and `pmax`. Calling `clamp(a, x, b)` computes the same result as `pmax(a, pmin(x, b))`. In other words, it returns the values of the vector x limited to a minimum value of a and a maximum value of b.

```
IntegerVector xx;
int a, b;

clamp( a, xx, b );
```
Listing 8.22 Example using `clamp` sugar function

8.3.2.15 unique

The `unique` function returns the subset of unique values among its input vector.

```
IntegerVector xx;

unique( xx );
```
Listing 8.23 Example using `unique` sugar function

8.3.2.16 sort_unique

The `sort_unique` function combines the results from `unique` with a call to `sort`.

8.3.2.17 table

The `table` function returns a named vector with counts of the occurrences of each of the named elements in the input vector, just like the R function `table`.

8.3 Functions

```
1  IntegerVector xx;

3  table( xx );
```

Listing 8.24 Example using `table` sugar function

8.3.2.18 duplicated

The `duplicated` function returns a logical vector indicated whether the element at position *i* in the input vector duplicates a previous value.

```
1  IntegerVector xx;

3  duplicated( xx );
```

Listing 8.25 Example using `duplicated` sugar function

8.3.3 Mathematical Functions

For the following set of functions, generally speaking, the *i*th element of the result of the given function (say, `abs`) is the result of applying that function to this *i*th element of the input expression. Supported types are integer and numeric.

Some functions reduce the input vector to a scalar result. Examples such as `min()`, `max()`, `mean()`, `var()` or `sd()` show that the commonly assumed functionality is also provided by these *Rcpp sugar* functions.

```
1  NumericVector x, y;
   int k;
3  double z;

5  abs(x);
   exp(x);
7  floor(x);
   ceil(x);
9  pow(x, z);      # x to the power of z
   log(x);
11 log10(x);
   sqrt(x);
13 sin(x);
   cos(x);
15 tan(x);
   sinh(x);
17 cosh(x);
   tanh(x);
19 asin(x);
   acos(x);
```

```
21  atan(x);
    gamma(x);
23  lgamma(x);   # log gamma
    digamma(x);
25  trigamma(x);
    tetragamma(x);
27  pentagamma(x);
    expm1(x);
29  log1p(x);
    factorial(x);
31  lfactorial(x);
    choose(n, k);
33  lchoose(n, k);
    beta(n, k);
35  lbeta(n, k);
    psigamma(n, k);
37  trunc(x);
    round(x, k);
39  signif(x, k);
    mean(x);
41  var(x);
    sd(x);
43  sum(x);
    cumsum(x);
45  min(x);
    max(x);
47  range(x);
    which_min(x);
49  which_max(x);
    setequal(x, y);
```

Listing 8.26 Examples using mathematical sugar functions

8.3.4 The d/q/p/q Statistical Functions

The framework provided by *Rcpp sugar* also permits easy and efficient access to the density, distribution function, quantile and random number generation functions used by R itself. These are also provided via the Rmath library and, as discussed in Sect. 4.8, available via the R namespace provided by the **Rcpp** package for this part of the R API.

In general, the functions provided by *Rcpp sugar* are vectorized for the first element. Consequently, in the following example, the function calls work in C++ just as they would in R:

```
  x1 = dnorm(y1, 0, 1);    // density of y1 at m=0, sd=1
2 x2 = pnorm(y2, 0, 1);    // distribution function of y2
  x3 = qnorm(y3, 0, 1);    // quantiles of y3
4 x4 = rnorm(n, 0, 1);     // 'n' RNG draws of N(0, 1)
```

Listing 8.27 Examples of d/p/q/r statistical sugar functions sugar

For x1 to x3, the resulting vector is of the same dimension as the input y1 to y3.

Similar d/q/p/r functions are provided for the most common distributions: beta, binom, cauchy, chisq, exp, F, gamma, geom, hyper, lnorm, logis, nbeta, nbinom, nbinom_mu, nchisq, nf, norm, nt, pois, t, unif, and weibull.

One important point is that the programmer using the random number generator functions needs to initialize the state of the random number generator as detailed in Section 6.3 of the "Writing R Extensions" manual (R Development Core Team 2012d). To help with this, the **Rcpp** package offers a convenient C++ solution: A *scoped* class that sets the random number generator on entry to a block and resets it on exit. The following example uses this RNGScope class. The function defines the code block in which the scoped variable is active; and here RNGScope is activated upon entering the function. Therefore, the random number generator can be called in order to assign values to x. The scoped variable is then destroyed after the core function code terminates with the return statement. RNGScope does not have to be the first statement. In fact, it can be placed anywhere in the function scope—but it has to be called before the first call to the random number generator is made.

```
RcppExport SEXP getRGamma() {
    RNGScope scope;
    NumericVector x = rgamma( 10, 1, 1 );
    return x;
}
```

Listing 8.28 Examples of using sugar RNG functions with RNGScope

As there is some computational overhead involved in using RNGScope, we are not wrapping it automatically around each inner function generating random numbers. Rather, the user of these functions should place an RNGScope at the appropriate level of his or her code.

In cases where scalar functions of a single argument returning a single are required, these are provided via R namespace as discussed in Sect. 4.8. The interface is identical to the one offered by the header file Rmath.h of the R installation. This additional interface is provided as *Rcpp sugar* requires the original header file to be used via a remapping through prefix Rf_, whereas the added functions can be used directly from the new namespace cleanly separating these identifiers.

8.4 Performance

The **Rcpp** package contains a complete example illustrating possible performance gains from using *Rcpp sugar* in the directory examples/SugarPerformance. It compares the performance on four different R expressions between calling running it as an R expression, running it via hand-optimized C++ code, and running it via the *Rcpp sugar* vectorized C++ approach. Four different expressions are evaluated covering any, ifelse (where we use two variants of the handwritten code with and without checks for missing values), and sapply.

Table 8.1 Run-time performance of *Rcpp sugar* compared to R and manually optimized C++

R expression	Runs	Manual	Sugar	R
any(x * y < 0)	5,000	0.00027	0.00069	6.8914
ifelse(x<y, x*x, -(y*y))	500	1.28566	1.52103	13.8829
ifelse(x<y, x*x, -(y*y)) (noNA)	500	0.41462	1.14434	13.8537
sapply(x, square)	500	0.16721	0.19224	115.4236

As can be seen in Table 8.1, performance varies greatly. This is most dramatic in the first example. An R expression such as any(x * y < 0) will always be evaluated for *all* pairwise elements in the two vectors. The *Rcpp sugar* implementation, on the other hand, can take a shortcut and stop as soon as one of the expressions swept across all pairwise elements evaluates to true. After all, the tests is only for "at least one" rather than a full count. The combination of compiled code together with a possible short-circuit exit makes the C++ implementation much faster. In fact, the ratio of C++ time to R time is almost 1–10,000. Handwritten code can still be faster than the *Rcpp sugar* code by a small factor.

The second and third examples illustrate the vector function introduced at the beginning of this chapter. We show two sets of results: In the second example, a standard implementation which, just like R itself, tests all elements for NA (which imposes an additional performance burden); and, the third example which does not perform these tests for NA. Here, we find a 9- and 12-fold gain, respectively, from *Rcpp sugar* relative to the vectorized R code. The manually written C++ code gains most from omitting the test for NA, whereas the default version is only marginally slower.

The fourth and final example illustrates sweeping a function (here computing a square of its argument) over a vector using sapply. Once again, the *Rcpp sugar* variant is a lot faster than the R version; the ratio of the two measurements is about 600. The *Rcpp sugar* code is only marginally slower than a handwritten loop.

This section illustrates that *Rcpp sugar* can offer substantial performance gains. While manually written C++ code is seen to be marginally faster, the more concise vectorized code offered by *Rcpp sugar* might be easier to write and maintain making it an attractive proposition.

8.5 Implementation

This section details some of the techniques used in the implementation of *Rcpp sugar*. Note that the user need not to be familiar with the implementation details in order to use *Rcpp sugar*, so this section can be skipped during an initial read of the chapter.

8.5 Implementation

Writing *Rcpp sugar* functions is fairly repetitive and follows a well-structured pattern. So once the basic concepts are mastered (which may take time given the inherent complexities in template programming), it should be possible to extend the set of functions further following the established pattern.

8.5.1 The Curiously Recurring Template Pattern

Expression templates such as those used by *Rcpp sugar* employ a technique called the *Curiously Recurring Template Pattern* (CRTP).[1] The general form of CRTP is:

```cpp
// The Curiously Recurring Template Pattern (CRTP)

// A templated base class
template <typename T>
struct base {
    // ...
};

// A derived class
// which is a template for the base class it inherits from
struct derived : base<derived> {
    // ...
};
```

Listing 8.29 The Curiously Recurring Template Pattern (CRTP)

The `base` class is templated by the class that derives from it: `derived`. This shifts the relationship between a base class and a derived class—and allows the base class to access methods of the derived class.

8.5.2 The VectorBase Class

The CRTP is used as the basis for *Rcpp sugar* with the `VectorBase` class template. All sugar expressions derive from one class generated by the `VectorBase` template. The current definition of `VectorBase` is given here:

```cpp
template <int RTYPE, bool na, typename VECTOR>
class VectorBase {
public:
    struct r_type :
        traits::integral_constant<int,RTYPE>{};
    struct can_have_na :
        traits::integral_constant<bool,na>{};
    typedef typename
```

---

[1] The Wikipedia page at http://en.wikipedia.org/wiki/Curiously_recurring_template_pattern has a good introduction and further pointers.

```
     traits::storage_type<RTYPE>::type stored_type;

     VECTOR& get_ref(){
         return static_cast<VECTOR&>(*this);
     }

     inline stored_type operator[]( int i) const {
         return static_cast<const VECTOR*>(this)->operator[](i);
     }

     inline int size() const {
         return static_cast<const VECTOR*>(this)->size();
     }

     /* definition omitted here */
     class iterator;

     inline iterator begin() const {
        return iterator(*this, 0);
     }
     inline iterator end() const {
        return iterator(*this, size() );
     }
}
```

Listing 8.30 The VectorBase class for *Rcpp sugar*

The VectorBase template has three parameters.

RTYPE   which controls the type of the underlying SEXP expression.

na   which embeds in the derived type information about whether instances may contain missing values. **Rcpp** vector types (IntegerVector,...) derive from VectorBase with this parameter set to true because there is no way to know at compile-time if the vector will contain missing values at run-time. However, this parameter is set to false for types that are generated by sugar expressions as these are guaranteed to produce expressions that are without missing values. An example is the is_na function. This parameter is used in several places as part of the compile time dispatch to limit the occurrence of redundant operations.

VECTOR   which is a key component of *Rcpp sugar*. This is the manifestation of CRTP. The indexing operator and the size method of VectorBase use a static cast of this to the VECTOR type to forward calls to the actual method of the derived class.

8.5.3 *Example:* sapply

As an example, the current implementation of sapply, supported by the template class Rcpp::sugar::Sapply is given below in Listing 8.31.

8.5 Implementation

```
template <int RTYPE, bool NA, typename T, typename Function>
class Sapply : public VectorBase<
  Rcpp::traits::r_sexptype_traits<
    typename ::Rcpp::traits::result_of<Function>::type >::rtype,
  true,
  Sapply<RTYPE,NA,T,Function>
  > {
public:
  typedef typename ::Rcpp::traits::result_of<Function>::type;

  const static int RESULT_R_TYPE =
    Rcpp::traits::r_sexptype_traits<result_type>::rtype;

  typedef Rcpp::VectorBase<RTYPE,NA,T> VEC;

  typedef typename Rcpp::traits::r_vector_element_converter<
      RESULT_R_TYPE>::type converter_type;

  typedef typename
    Rcpp::traits::storage_type< RESULT_R_TYPE>::type STORAGE;

  Sapply( const VEC& vec_, Function fun_ ) :
    vec(vec_), fun(fun_){}

  inline STORAGE operator[]( int i ) const {
    return converter_type::get( fun( vec[i] ) );
  }

  inline int size() const { return vec.size(); }

private:
    const VEC& vec;
    Function fun;
};

// sugar

template <int RTYPE, bool _NA_, typename T, typename Function>
inline sugar::Sapply<RTYPE,_NA_,T,Function>
sapply(const Rcpp::VectorBase<RTYPE,_NA_,T>& t, Function fun) {
    return sugar::Sapply<RTYPE,_NA_, T,Function>(t, fun);
}
```

Listing 8.31 The sapply *Rcpp sugar* implementation

8.5.3.1 The sapply Function

`sapply` is a template function that takes two arguments.

- The first argument is a sugar expression, which we recognize because of the relationship with the `VectorBase` class template.
- The second argument is the function `fun` to apply.

The `sapply` function itself does not do anything. Rather, it is used to trigger compiler detection of the template parameters that will be used in the `sugar::Sapply` template.

8.5.3.2 Detection of Return Type of the Function

In order to decide what kind of expression is built, the `Sapply` template class queries the template argument via the `Rcpp::traits::result_of` template.

```
typedef typename
  ::Rcpp::traits::result_of<Function>::type result_type;
```

Listing 8.32 `Rcpp::traits::result_of` template

The `result_of` type trait is implemented as follows:

```
template <typename T>
struct result_of {
  typedef typename T::result_type type;
};

template <typename RESULT_TYPE, typename INPUT_TYPE>
struct result_of< RESULT_TYPE (*)(INPUT_TYPE) >{
    typedef RESULT_TYPE type;
};
```

Listing 8.33 `result_of` trait implementation

The generic definition of `result_of` targets functors which contain a nested `result_type` type. The second definition is a partial specialization targeting function pointers.

8.5.3.3 Identification of Expression Type

Based on the result type of the function, the `r_sexptype_traits` trait is used to identify the expression type.

```
const static int RESULT_R_TYPE =
  Rcpp::traits::r_sexptype_traits<result_type>::rtype;
```

Listing 8.34 `Rcpp::traits::r_sexptype_traits` template

8.5 Implementation

8.5.3.4 Converter

The `r_vector_element_converter` class is used to convert an object of the function's result type to the actual storage type suitable for the sugar expression.

```
  typedef typename
2   Rcpp::traits::r_vector_element_converter<RESULT_R_TYPE>::type
    converter_type;
```

Listing 8.35 `r_vector_element_converter` class

8.5.3.5 Storage Type

The `storage_type` trait is used to get access to the storage type associated with a sugar expression type. For example, the storage type of a `REALSXP` expression is `double`.

```
1 typedef typename
    Rcpp::traits::storage_type<RESULT_R_TYPE>::type STORAGE;
```

Listing 8.36 `storage_type` trait

8.5.3.6 Input Expression Base Type

The input expression—the expression over which `sapply` runs—is also defined in a `typedef` for convenience:

```
typedef Rcpp::VectorBase<RTYPE,NA,T> VEC;
```

Listing 8.37 Input expression base type

8.5.3.7 Output Expression Base Type

In order to be part of the *Rcpp sugar* system, the type generated by the `Sapply` class template must inherit from `VectorBase`.

```
1 template <int RTYPE, bool NA, typename T, typename Function>
  class Sapply : public VectorBase<
3   Rcpp::traits::r_sexptype_traits<
      typename ::Rcpp::traits::result_of<Function>::type>::rtype,
5   true,
    Sapply<RTYPE, NA, T, Function>
7 >
```

Listing 8.38 Output expression base type

Here we have three arguments. First, the expression built by `Sapply` depends on the result type of the function. Second, it may contain missing values. The third argument is the manifestation of the *CRTP*.

8.5.3.8 Constructor

The constructor of the `Sapply` class template is straightforward, it simply consists of holding the reference to the input expression and the function.

```
Sapply( const VEC& vec_, Function fun_ ) :
  vec(vec_), fun(fun_){}

private:
  const VEC& vec;
  Function fun;
```
Listing 8.39 Constructor for `Sapply` class template

8.5.3.9 Implementation

The indexing operator and the `size` member function is what the `VectorBase` expects. The size of the result expression is the same as the size of the input expression and the *i*th element of the result is simply retrieved by applying the function and the converter. Both these methods are inlined to maximize performance:

```
inline STORAGE operator[]( int i ) const {
  return converter_type::get( fun( vec[i] ));
}
inline int size() const { return vec.size(); }
```
Listing 8.40 Implementation of `Sapply`

8.6 Case Study: Computing π Using *Rcpp sugar*

Rcpp sugar provides a large number of functions that can be used as building blocks for other programs and applications. Rather than picking a particular example from an existing package, this section will demonstrate how *Rcpp sugar* is as compact and expressive as R itself.

To do so, we will revisit the well-known introductory example of approximating π. The algorithm uses the property that the area of a unit circle is equal to π and repeatedly draws two uniform random numbers x and y, each between zero and one. It then computes the distance

$$d = \sqrt{x^2 + y^2}$$

8.6 Case Study: Computing π Using *Rcpp sugar*

to the origin and compares it to one to determine whether the point (x,y) is inside or outside the unit circle. By summing up all the attempts less than one, dividing by the total count N, one obtains a proportion—of the area of a quarter of the unit circle as we constrained the initial draws to be in the first quadrant. So, consequently, our estimate of π is provided by four times that area as an estimate for the area of the whole unit circle.

```
piR <- function(N) {
    x <- runif(N)
    y <- runif(N)
    d <- sqrt(x^2 + y^2)
    return(4 * sum(d < 1.0) / N)
}
```

Listing 8.41 Simulating π in R

An equivalent program can be written in C++ thanks to *Rcpp sugar* with only one more line (in the function body) to ensure the random number generator is properly set.

```
#include <Rcpp.h>

using namespace Rcpp;

// [[Rcpp::export]]
double piSugar(const int N) {
    RNGScope scope;              // ensure RNG gets set/reset
    NumericVector x = runif(N);
    NumericVector y = runif(N);
    NumericVector d = sqrt(x*x + y*y);
    return 4.0 * sum(d < 1.0) / N;
}
```

Listing 8.42 Simulating π in C++

Using Rcpp attributes, we can obtain an R function of the same name `piSugar` simply by passing the name of the file into the `sourceCpp()` function.

The complete example is shown in Listing 8.43.

```
library(Rcpp)
library(rbenchmark)

piR <- function(N) {
    x <- runif(N)
    y <- runif(N)
    d <- sqrt(x^2 + y^2)
    return(4 * sum(d < 1.0) / N)
}

# get C++ version from source file
sourceCpp("piSugar.cpp")

N <- 1e6
```

```
16  set.seed(42)
    resR <- piR(N)
18
    set.seed(42)
20  resCpp <- piSugar(N)

22  ## important: check results are identical with RNG seeded
    stopifnot(identical(resR, resCpp))
24
    res <- benchmark(piR(N), piSugar(N), order="relative")
26
    print(res[,1:4])
```

Listing 8.43 Simulating π in R

The result is shown in Table 8.2. Even though both versions are equally compact and execute essentially identical vectorized code leading to identical results (as we verified), the C++ version manages to reduce the run-time by almost half which is a surprisingly good result.

Table 8.2 Run-time performance of *Rcpp sugar* compared to R for simulating π

R expression	Replications	Elapsed	Relative
piSugar(N)	100	5.777	1.000
piR(N)	100	11.227	1.943

We should stress, though, that one should not take away from this to attempt to rewrite each and every R expression using *Rcpp sugar*. Its power lies in providing us concise expression at the C++ level. This allows the programmer to write compact code similar to what one could achieve in R. This can complement other C++ code and is not necessarily meant to replace R as vectorized R code is typically fast enough.

Part IV
Applications

Chapter 9
RInside

Abstract The **RInside** package permits direct use of R inside of a C++ application. **RInside** provides an abstraction layer around the R embedding API and makes it easier to access an R instance inside your application. Moreover, thanks to the classes provided by **Rcpp**, data interchange between R and C++ becomes very straightforward. We illustrate **RInside** by examining several of the many examples included with the package.

9.1 Motivation

Rcpp has been discussed throughout this book as a package which makes it easier to add new code to R itself. Examples for this can be stand-alone routines, accessing an external library or a combination of both. One commonality has been that R remains the principal interface: The aim is always to extend R with new facilities, yet the focus is on extending R as the principal environment for statistical computing, data analysis, and modeling.

However, there are situations where one may want to take a different view, starting from a C++ program as a main executable. One might like to deploy R as an analysis engine or service to enhance the executable. As a more concrete example, consider the case of a (potentially large) program to control a set of simulations. After running a number of experiments, results can be aggregated to be analyzed further with the intent of deriving intermediate results which will influence further simulations. Our quest is to make the intermediate analysis accessible to the outer C++ program controlling the simulation. So at this point, one of the two common workflows is usually deployed.

The first approach is the simplest. Data is written out to files. Standard textfiles are common, or maybe a domain-specific or higher-performance binary format is chosen. At this point, analysis may switch to another program for data analysis such as R. The analysis itself maybe written out in scripts; maybe these can be executed with a front-end such as `Rscript` (which is available wherever R is installed).

And, in that case the main analysis program could even call `Rscript` itself via a `system()` call. Once the data analysis has completed, the main simulation program can resume, possibly reflecting updated parameters from the analysis.

A second approach may be to communicate with the analysis program over the network. The **Rserve** package (Urbanek 2003, 2012) can listen on network sockets and receive data as well as instructions—which makes it suitable to act as a networked R analysis facility. The main program can transmit the data to be analyzed and can then invoke the analysis script. The main program proceeds with its operations once it receives the results from the analysis engine or server. Thus, this setup loosely corresponds to the facilities from the first scripting case in that calls are made to an external engine.

Both approaches work, yet both also have drawbacks. Calling an external program using `system()` is relatively simple and somewhat robust. However, a key problem is the difficulty of reporting errors back via the reduced interface offered by the `system()` call. One can encode the respective success and error codes as integer values which can be returned. Alternatively, results can be written out to the standard output, or into files, which the main program then needs to parse. No other formal mechanism exists for communication between the two programs sharing the work. Another drawback concerning file-based communication is the possibility of race conditions if more than one instance is running. Similarly, the networked solution introduces a possible new point of failure at the network layer. This can of course be mitigated by running the **Rserve** instance on the same machine, or by adding some redundancy to the network setup.

The potential shortcomings of both approaches suggest a search for alternatives. One possibility supported by R is to *embed* the interpreter in another program. This is supported through an extensive embedding API (R Development Core Team 2012d, Chapter 8). This API is written in C and does not support higher-level abstractions as in **Rcpp**. However, the **RInside** package (Eddelbuettel and François 2012d) builds on top of the standard R embedding API in a way that may seem more natural to C++ programmers. In this chapter, we illustrate the use of **RInside** via a number of examples supplied with the package itself.

9.2 A First Example: Hello, World!

Let us look at the very first of well over a dozen examples presented in the directory `examples/standard` of the **RInside** package. It follows the long and proud tradition of making a first program display the string "Hello, world!" on the screen:

```
#include <RInside.h>

int main(int argc, char *argv[]) {

    // create an embedded R instance
    RInside R(argc, argv);

```

9.2 A First Example: Hello, World!

```
       // assign a char* (string) to 'txt'
 9     R["txt"] = "Hello, world!\n";

11     // eval the init string, ignoring any returns
       R.parseEvalQ("cat(txt)");
13
       exit(0);
15 }
```

Listing 9.1 First **RInside** example: Hello, World!

The program really consists of only four statements, and one single header file (RInside.h) providing all declarations. We first instantiate an object called R of the RInside class. The class has two arguments to deal with command-line arguments; these arguments are, however, entirely optional. This is followed by an assignment of a constant text string—the message to be displayed—to a variable named txt which is created directly inside the R session. Next, we parse and evaluate an R command passed to the embedded R instances which calls the cat() function to simply display the content of variable txt. This last parse and evaluation is done "quietly" (as indicated by the trailing "Q" on the member function name) and no result is returned. The related function parseEval() which we will see below returns the value of its last expression, much like a standard R function. Finally, we return with error code of zero, which is a common value to indicate successful completion.

Building this first program is straightforward provided that a Makefile (or Makefile.win for the Windows platform) has been set up—as is the case with the aforementioned directory examples/standard of the **RInside** package. The Makefile contains shell expressions which query R, **Rcpp**, and **RInside** for relevant header and library information, and then use this information to build the complete compile command:

```
 1 ## comment this out if you need a different version of R,
   ## and set set R_HOME accordingly as an environment variable
 3 R_HOME :=        $(shell R RHOME)

 5 sources :=       $(wildcard *.cpp)
   programs :=     $(sources:.cpp=)

 7
 9 ## include headers and libraries for R
   RCPPFLAGS :=    $(shell $(R_HOME)/bin/R CMD config --cppflags)
11 RLDFLAGS :=     $(shell $(R_HOME)/bin/R CMD config --ldflags)
   RBLAS :=        $(shell $(R_HOME)/bin/R CMD config BLAS_LIBS)
13 RLAPACK :=      $(shell $(R_HOME)/bin/R CMD config LAPACK_LIBS)

15 ## if you need to set an rpath to R itself, also uncomment
   #RRPATH :=      -Wl,-rpath,$(R_HOME)/lib
17
   ## include headers and libraries for Rcpp interface classes
19 RCPPINCL :=     $(shell echo 'Rcpp:::CxxFlags()' | \
                   $(R_HOME)/bin/R --vanilla --slave)
```

```
21 RCPPLIBS  :=      $(shell echo 'Rcpp:::LdFlags()' | \
                     $(R_HOME)/bin/R --vanilla --slave)
23
   ## include headers and libraries for RInside embedding classes
25 RINSIDEINCL :=    $(shell echo 'RInside:::CxxFlags()' | \
                     $(R_HOME)/bin/R --vanilla --slave)
27 RINSIDELIBS :=    $(shell echo 'RInside:::LdFlags()' | \
                     $(R_HOME)/bin/R --vanilla --slave)
29
   ## compiler etc settings used in default make rules
31 CXX       :=      $(shell $(R_HOME)/bin/R CMD config CXX)
   CPPFLAGS  :=      -Wall \
33                   $(shell $(R_HOME)/bin/R CMD config CPPFLAGS)
   CXXFLAGS  :=      $(RCPPFLAGS) $(RCPPINCL) $(RINSIDEINCL) \
35                   '$(R_HOME)"/bin/R CMD config CXXFLAGS'
   LDLIBS    :=      $(RLDFLAGS) $(RRPATH) $(RBLAS) $(RLAPACK) \
37                   $(RCPPLIBS) $(RINSIDELIBS)
```

Listing 9.2 Makefile for **RInside** examples

With a Makefile in place, we merely have to say make rinside_sample0 to build this first example, or even just make to build all examples. The equivalent manual build commands are displayed below as the output from calling make. The exact form will differ depending on where packages have been installed as well as operating system-specific aspects and local system-wide compiler flags. To provide an illustration for a Linux system, we see the following execute (with lines broken for display purposes):

```
1 sh> make rinside_sample0
  g++ -I/usr/share/R/include
3    -I/usr/local/lib/R/site-library/Rcpp/include
     -I"/usr/local/lib/R/site-library/RInside/include"
5    -O3 -pipe -g -Wall rinside_sample0.cpp
     -L/usr/lib64/R/lib -lR   -lblas -llapack
7    -L/usr/local/lib/R/site-library/Rcpp/lib -lRcpp
     -Wl,-rpath,/usr/local/lib/R/site-library/Rcpp/lib
9    -L/usr/local/lib/R/site-library/RInside/lib -lRInside
     -Wl,-rpath,/usr/local/lib/R/site-library/RInside/lib
11   -o rinside_sample0
```

Listing 9.3 Using Makefile for **RInside** to build example

This multiline expression looks somewhat intimidating. But it really only combines three sets of headers and libraries. These come from three distinct sources which are combined into one larger compile-and-link command:

1. For compiling and linking with R as would be provided by R CMD COMPILE and R CMD LINK.
2. For compiling and linking with **Rcpp**.
3. For compiling and linking with **RInside**.

Users can simply copy the same base components from the provided Makefile to build their own Makefile. As an alternative, the provided Makefile is generic and can be reused. It will compile and link any collection of example files into the corresponding set of executable programs. The only (current) constraint is the one-to-one mapping between source files and executables. Multiple dependencies are not currently supported though this could be added by extending the Makefile.

Finally, on the Windows platform the corresponding Windows Makefile has to be used. This can be accomplished most easily by supplying the -f Makefile.win argument to the make invocation as in make -f Makefile.win.

9.3 A Second Example: Data Transfer

Several other examples are supplied with the **RInside** package in a directory examples/standard. Several of these examples contain simple usage cases of calling R functions. The example below, which is a lightly edited version of the file rinside_sample6.cpp, shows simple data transfers for a variety of containers around double types:

```
#include <RInside.h>

int main(int argc, char *argv[]) {

    RInside R(argc, argv);

    double d1 = 1.234;              // scalar double
    R["d1"] = d1;

    std::vector<double> d2;         // vector of doubles
    d2.push_back(1.23);
    d2.push_back(4.56);
    R["d2"] = d2;

    std::map< std::string, double > d3; // map
    d3["a"] = 7.89;
    d3["b"] = 7.07;
    R["d3"] = d3;

    std::list< double > d4;         // list of doubles
    d4.push_back(1.11);
    d4.push_back(4.44);
    R["d4"] = d4;

    std::string txt =               // now access in R
        "cat('\nd1=', d1, '\n'); print(class(d1));"
        "cat('\nd2=\n');print(d2);print(class(d2));"
        "cat('\nd3=\n');print(d3);print(class(d3));"
        "cat('\nd4=\n');print(d4);print(class(d4));";
```

```
        R.parseEvalQ(txt);
31
        exit(0);
33  }
```

Listing 9.4 Second **RInside** example: data transfer

This shows how to transfer, respectively, a single scalar floating-point number in double precision as well as the STL containers vector, map, and list for such floating-point numbers. Integers, characters, and logical could be passed through accordingly.

9.4 A Third Example: Evaluating R Expressions

The third example (based on `rinside_sample7.cpp`) shows that accessing results from a normal evaluation (as opposed to "quiet" as in the first example) in R is also very straightforward. We can display the results at the C++ side using C++ output operators as **Rcpp** has taken care of all transfers.

```
1  #include <RInside.h>

3  int main(int argc, char *argv[]) {

5      RInside R(argc, argv);

7      // assignment can be done directly via []
       R["x"] = 10 ;
9      R["y"] = 20 ;

11     // R statement evaluation and result
       R.parseEvalQ("z <- x + y");
13
       // retrieval access using [] and implicit wrapper
15     int sum = R["z"];
       std::cout << "10 + 20 = " << sum << std::endl ;
17
       // we can also return the value directly
19     sum = R.parseEval("x + y") ;
       std::cout << "10 + 20 = " << sum << std::endl ;
21
       exit(0);
23 }
```

Listing 9.5 Third **RInside** example: data transfer

This third example only shows a transfer of scalar values. However, larger composite objects can also be returned due to the implicit use of `wrap()`.

This last aspect is very important as any expression in R is represented as a SEXP, and as such SEXP objects can be transferred between R and C++ with ease thanks to the **Rcpp** functions `as<>()` and `wrap()`. By using these facilities, we actually have a perfectly generic and extensible way of passing vectors, matrices, data frames, and lists—or even combinations of these types—simply by relying on the templated code in **Rcpp**.

9.5 A Fourth Example: Plotting from C++ via R

The fourth and last example for **RInside** in this section is based on the file rinside_sample8.cpp. It shows that one can call the R function plot() from C++ function as well.

```
#include <RInside.h>
#include <unistd.h>

int main(int argc, char *argv[]) {

    // create an embedded R instance
    RInside R(argc, argv);

    // evaluate an R expression with curve()
    // because RInside defaults to interactive=false we use a file
    std::string cmd = "tmpf <- tempfile('curve'); "
        "png(tmpf); "
        "curve(x^2, -10, 10, 200); "
        "dev.off();"
        "tmpf";
    // we get the last assignment back, here the filename
    std::string tmpfile = R.parseEval(cmd);

    std::cout << "Could now use plot in " << tmpfile << std::endl;
    unlink(tmpfile.c_str());       // cleaning up

    // alternatively, by forcing a display we can plot to screen
    cmd = "x11(); curve(x^2, -10, 10, 200); Sys.sleep(30);";
    R.parseEvalQ(cmd);      // parseEvalQ evals without assignment

    exit(0);
}
```

Listing 9.6 Fourth **RInside** example: plotting from C++ via R

A small set of R instructions selects a temporary file to be used for Portable Network Graphics (PNG) file. A simple curve is then plotted. Here, we remove the temporary file but its name and location could be passed to another function displaying it.

For completeness, we also plot onto a normal graphics device. This assumes that such a device can be opened as in normal interactive mode. As this example shows, the embedded R instance is capable of executing the same set of instructions as an interactive R session.

9.6 A Fifth Example: Using RInside Inside MPI

Besides the directory examples/standard containing the example discussed so far in this chapter, the **RInside** package contains a directory examples/mpi which shows how to use R and **Rcpp** in the context of the Message Passing Interface (MPI). MPI is a very mature standard used particularly in scientific computing. It enables clusters of computers to work concurrently on programming problems. A detailed discussion of MPI is far beyond the scope of this sections; standard references exist (Gropp et al. 1996, 1999).

A simple example rinside_mpi_sample2.cpp follows. It is based on an earlier version which was kindly provided by Jianping Hua and which used the C version of the MPI standard. We have updated it to the C++ variant of the MPI API; both API variants are rather close.

```
#include <mpi.h>       // mpi header
#include <RInside.h>   // for the embedded R via RInside

int main(int argc, char *argv[]) {

    // mpi initialization
    MPI::Init(argc, argv);
    // obtain current node rank and total nodes running
    int myrank = MPI::COMM_WORLD.Get_rank();
    int nodesize = MPI::COMM_WORLD.Get_size();

    // create an embedded R instance
    RInside R(argc, argv);

    std::stringstream txt;
    // node information
    txt << "Hello from node " << myrank
        << " of " << nodesize << " nodes!" << std::endl;
    // assign string var to R variable 'txt'
    R.assign( txt.str(), "txt");

    // show node information
    std::string evalstr = "cat(txt)";
    // eval the init string, ignoring any returns
    R.parseEvalQ(evalstr);

    // mpi finalization
    MPI::Finalize();

    exit(0);
}
```

Listing 9.7 Fifth **RInside** example: parallel computing with MPI

This program simply prints the standard "Hello, World!" greeting from each of the nodes in an MPI cluster. It needs to build against MPI headers and libraries; the supplied Makefile does that for the Open MPI standard implementation. It can easily be adapted to different local deployments as well.

A somewhat richer example `rinside_mpi_sample3.cpp` is available as well and shows how to do some simple computations on each node in the MPI cluster.

9.7 Other Examples

The `examples/standard` directory contains a number of additional examples which may be of interest. Among the topics illustrated are:
- How to pass two-dimensional data structures such as matrices.
- Running a regression in R and displaying the results via C++ indicating how to deploy R as a backend for a C++ application.
- A small portfolio management problem motivated a mailing list post.
- Conversions examples for logicals, lists, and tests for environments.
- An example demonstrating how to use *Rcpp modules* with **RInside**.

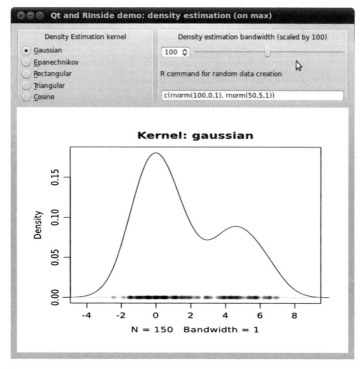

Fig. 9.1 Combining **RInside** with the Qt toolkit for a GUI application

The `examples/qt` directory shows an example of how to embed R inside of an application using the powerful and popular **Qt** toolkit supporting cross-platform applications, and in particular those with a graphical user interface (GUI) (Fig. 9.1).

This example is fairly straightforward. Given a mixture distribution (for which the user can alter parameter as well as the functional form), the choice of kernel function and parameters for the density estimation directly influence the density estimate. The application provides an opportunity to interactively experiment with these choices. Several of the so-called widgets in the GUI toolkit return parameters. For the selection of the estimation kernel, as well as the estimation bandwidth, this is an integer. For the R expression denoting the generation of the data set from which the density is estimated, we obtain a character string. These can be passed to R rather easily using the **Rcpp** facilities we have studied.

From R, we then obtain an updated graphics file and the GUI toolkit covers displaying the updated file. The remainder of the program—which is relatively short at about two hundred lines—mostly deals with the typical code to set up a graphical application, lay out the GUI widgets, and organize the callback events. Adding R into the mix is straightforward, thanks to RInside.

Similarly, the examples/wt directory shows two examples which implement a web application (corresponding to the GUI application written using Qt) by using the **Wt** ("web toolkit") library. **Wt** is responsible for all aspects of the network programming, provides an integrated webserver, and negotiates the best communications protocol with the webclient sending the request (Fig. 9.2).

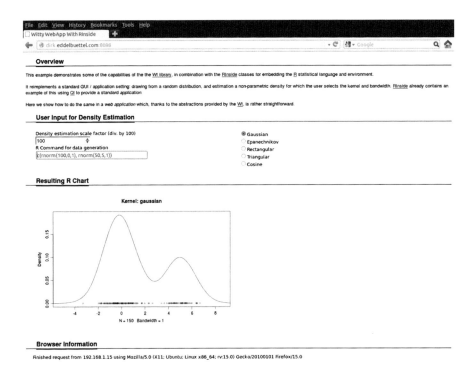

Fig. 9.2 Combining **RInside** with the Wt toolkit for a web application

9.7 Other Examples

The web application example is very similar to the GUI application as the Web toolkit library **Wt** covers all aspects of the network communication. Similar to the **Qt** example, the programmer simply has to receive the user choices on user events and update the estimated density based on these choices. This example uses cascading style sheets (CSS) to allow alteration of the appearances of the application without requiring any code logic and supports an XML file containing widget labels permitting similar changes to the descriptive texts without requiring an application rebuild.

Chapter 10
RcppArmadillo

Abstract The **RcppArmadillo** package implements an easy-to-use interface to the **Armadillo** library. **Armadillo** is an excellent, modern, high-level C++ library aiming to be as expressive to use as a scripting language while offering high-performance code due to modern C++ design including template meta- programming. **RcppArmadillo** brings all these features to the R environment by leaning on the **Rcpp** interface. This chapter introduces **Armadillo** and provides a motivating example via a faster replacement function for fitting linear models before it discusses a detailed case study of implementing a Kalman filter in **RcppArmadillo**.

10.1 Overview

Armadillo (Sanderson 2010) is a modern C++ library with a focus on linear algebra and related operations. Quoting from its homepage:

> **Armadillo** is a C++ linear algebra library aiming towards a good balance between speed and ease of use. The syntax is deliberately similar to Matlab. Integer, floating point and complex numbers are supported, as well as a subset of trigonometric and statistics functions. Various matrix decompositions are provided [...].
>
> A delayed evaluation approach is employed (during compile time) to combine several operations into one and reduce (or eliminate) the need for temporaries. This is accomplished through recursive templates and template meta-programming.
>
> This library is useful if C++ has been decided as the language of choice (due to speed and/or integration capabilities) [...].

Its features can be illustrated with a simple example from the **Armadillo** web site (which we modified slightly to fit with the style of the rest of the book).

```
#include <iostream>
#include <armadillo>

using namespace std;
using namespace arma;
```

```
7 int main(int argc, char** argv) {
     mat A = randn<mat>(4,5);
9    mat B = randn<mat>(4,5);

11   cout << A * trans(B) << endl;

13   return 0;
  }
```

Listing 10.1 A simple **Armadillo** example

Header files for both the standard input/output streams and **Armadillo** itself are included to provide the required declarations. Two **Armadillo** matrices A and B of size 4×5 are then filled with random variables drawn from a $N(0, 1)$ distribution. The randn() function is templated to the matrix type. Finally, the 4×4 matrix resulting from multiplying A by the transpose of B is computed and the result is printed to the standard output.

The code is highly readable and easy to study. This is in one part due to the global namespace import for both std and arma which shortens the function and class names. It is also due to the sensible use of identifiers such as trans(B) for a transpose (and the alternate form B.t() is also supported), as well as mat as a default matrix type. This is in fact a typedef for Mat<double>, a matrix templated to the standard floating-point type. Other matrix and vector types exist for integers, unsigned integers, and complex numbers.

Armadillo supports a large number of functions as a look at its available documentation reveals. While many of these functions are also available within R itself, they make **Armadillo** as attractive choice for the C++ programmer aiming to easily extend functionality at the C++ source level. This, in essence, is the main attraction of **Armadillo**: an easy-to-use, feature-complete, well-supported modern C++ library for linear algebra. The **RcppArmadillo** package (François et al. 2012; Eddelbuettel and Sanderson 2013) integrates it into R using facilities provided by the **Rcpp** package.

10.2 Motivation: FastLm

10.2.1 Implementation

Fitting linear models is a fundamental building block of data analysis. It is available in R via the powerful lm() function which provides a vast amount of additional functionality, as well as the more spartan lm.fit() function.

Below, we show the complete file fastLm.cpp from the src directory of the **RcppArmadillo** package which implements a faster replacement function suitable for use in extended simulations.

10.2 Motivation: FastLm

```cpp
extern "C" SEXP fastLm(SEXP ys, SEXP Xs) {

    try {
        // Rcpp and arma structure reuse original memory
        Rcpp::NumericVector yr(ys);
        Rcpp::NumericMatrix Xr(Xs);
        int n = Xr.nrow(), k = Xr.ncol();
        arma::mat X(Xr.begin(), n, k, false);
        arma::colvec y(yr.begin(), yr.size(), false);
        int df = n - k;

        // fit model y ~ X, extract residuals
        arma::colvec coef = arma::solve(X, y);
        arma::colvec res  = y - X*coef;

        double s2 = std::inner_product(res.begin(), res.end(),
                                       res.begin(), 0.0)/df;
        // std.errors of coefficients
        arma::colvec sderr = arma::sqrt(s2 *
            arma::diagvec(arma::pinv(arma::trans(X)*X)));

        return Rcpp::List::create(Rcpp::Named("coefficients")=coef,
                                  Rcpp::Named("stderr")      =sderr,
                                  Rcpp::Named("df")          =df);

    } catch( std::exception &ex ) {
        forward_exception_to_r( ex );
    } catch(...) {
        ::Rf_error( "c++ exception (unknown reason)" );
    }
    return R_NilValue; // -Wall
}
```

Listing 10.2 FastLm function using **RcppArmadillo**

As the example demonstrates, **Armadillo** allows us to write remarkably compact code:

1. We start by instantiating **Rcpp** objects for the model matrix and dependent variable; these are lightweight proxy objects and no data is copied.
2. The vector y and matrix X are initialized as an arma matrix and vector from the **Rcpp** types using the dimension information and iterator pointing to the beginning of the data, and again no explicit memory allocation is needed.
3. The model fit of $y \sim X$ is also done in one solve() statement, as is the calculation of the residuals as $y - X\hat{\beta}$.
4. Similarly, the sum of squared residuals is computed in a single statement, thanks to the STL inner_product function, and the result is divided by the degrees of freedom $n - k$.
5. Now the standard errors of the estimate are extracted as the squared root of the diagonal of $(X'X)^{-1}$, scaled by the sum of squared residuals.
6. We need one further statement to create the named list of return values.

7. Controlling for exceptions is straightforward with a `try/catch` block which passes recognized exception back to R using a helper function, or shows a default text in case of an unrecognized exception.

The **RcppArmadillo** package provides access to the function above via two different interfaces. The simpler function `fastLmPure()` just transfers a given vector and matrix and executes the regression, without any other transformation (but two tests for suitable data type and conforming dimensions). The higher-level function `fastLm()` provides the standard modeling interface using the common formula notation.

10.2.2 Performance Comparison

As before, we can rely on **inline** to create a function by compiling, linking, and loading the code below. This is made possible by a plugin provided by the **RcppArmadillo** package and used by the **inline** to determine the required values to instrument the underlying R CMD COMPILE and R CMD SHLIB calls which execute the build.

```
src <- '
    Rcpp::NumericMatrix Xr(Xs);
    Rcpp::NumericVector yr(ys);
    int n = Xr.nrow(), k = Xr.ncol();
    arma::mat X(Xr.begin(), n, k, false);
    arma::colvec y(yr.begin(), yr.size(), false);
    int df = n - k;

    // fit model y ~ X, extract residuals
    arma::colvec coef = arma::solve(X, y);
    arma::colvec res  = y - X*coef;

    double s2 = std::inner_product(res.begin(), res.end(),
                                    res.begin(), 0.0)/df;
    // std.errors of coefficients
    arma::colvec sderr = arma::sqrt(s2 *
        arma::diagvec(arma::pinv(arma::trans(X)*X)));

    return Rcpp::List::create(Rcpp::Named("coefficients")=coef,
                              Rcpp::Named("stderr")        =sderr,
                              Rcpp::Named("df")            =df);
'
fLm <- cxxfunction(signature(Xs="numeric", ys="numeric"),
                   src, plugin="RcppArmadillo")
```

Listing 10.3 Basic `fLm()` function without formula interface

We can also time and compare the approaches. As before, we use the **rbenchmark** package (Kusnierczyk 2012) which contains a function `benchmark()` which makes such timing comparisons very straightforward. We use the data set

10.2 Motivation: FastLm

trees which is also used in the example from the help page for the original lm() function in R and compare computation of the linear fit via the three different approaches, each repeated one thousand times.

As intermediate approaches, we use both the fastLmPure() function implemented in **RcppArmadillo** and a simpler fastLmPure2() used just here. We can think of fastLmPure() as an equivalent to lm.fit() as it works directly on a matrix and vector rather than a model formula. The function is defined as follows:

```
fastLmPure <- function(X, y) {

    stopifnot(is.matrix(X))
    stopifnot(nrow(y)==nrow(X))

    .Call("fastLm", X, y, PACKAGE = "RcppArmadillo")
}
```

Listing 10.4 Basic fastLmPure() R function without formula interface

The fastLmPure2() is a copy where we removed the two stopifnot() tests for proper types and dimensions. That is done just for this performance comparison and not recommended for normal production code.

Given these functions, the small performance comparison can be executed as follows:

```
R> y <- log(trees$Volume)
R> X <- cbind(1, log(trees$Girth))
R> frm <- formula(log(Volume) ~ log(Girth))
R> benchmark(fLm(X, y),
+            fastLmPure(X, y),
+            fastLmPure2(X, y),
+            fastLm(frm, data=trees),
+            columns = c("test", "replications",
+                        "elapsed", "relative"),
+            order="relative",
+            replications=1000)

                     test replications elapsed   relative
1              fLm(X, y)          1000   0.034   1.000000
3      fastLmPure2(X, y)          1000   0.040   1.176471
2       fastLmPure(X, y)          1000   0.081   2.382353
5            lm.fit(X, y)         1000   0.136   4.000000
4 fastLm(frm, data = trees)      1000   1.414  41.588235
```

Listing 10.5 FastLm comparison

Given the small size of the data set, executing the underlying regression is not expensive at all. Hence, small code differences such as the testing for data type and data dimension (which is done by fastLmPure() but not by fastLmPure2()) can have a disproportionate performance impact. It may appear magnified here given the short computation time required by the minimal implementation in fLm() from the cxxfunction() invocation above.

The more dramatic difference is between the formula interface offered by the **RcppArmadillo** package and the more direct implementation. The additional function calls needed to parse the formula and to set up the model matrix can be seen as having a disproportionate cost especially given the small size of the data set used here. We also see that fLm(), with its directly attached object code and pointer to it, performs marginally faster than the simplest possible implementation in a function using .Call() employing code from the indicated package, here **RcppArmadillo**. However, this small difference is dwarfed by the cost of the (highly recommended) tests for proper types and dimensions in fastLmPure().

Overall the example is very encouraging. We can execute one thousand calls to the simple regression function created on the fly via the **inline** package in about 34 ms. The bare bones implementation of fastLmPure from **RcppArmadillo** takes a little longer at 81 ms. By reducing fastLmPure further to just the .Call() invocation, its time moves within twenty percent of the time of fLm which indicates a small but noticeable overhead from the .Call() interface.

10.2.3 A Caveat

The reimplementation of lm() using **Armadillo** has served as a very useful example of how to add C++ code implementing linear algebra operations. However, there is one important difference between the *numerical computing* aspect and the *statistical computing* side. The help page for fastLm in the **RcppArmadillo** package has an illustration. *Numerical computing Statistical computing*

It uses an artificial data set constructed such that it produces a rank-deficient two-way layout with missing cells. Such cases require a special pivoting scheme of "pivot only on (apparent) rank deficiency" which R contains via customized routines based on the **Linpack** library. This provides the appropriate statistical computing approach, but such pivoting is generally not contained in any conventional linear algebra software libraries such as **Armadillo**.

```
R> ## case where fastLm breaks down
R> dd <- data.frame(f1 = gl(4, 6, labels=LETTERS[1:4]),
+                   f2 = gl(3, 2, labels=letters[1:3]))[-(7:8),]
R> xtabs(~ f2 + f1, dd)     # one missing cell
   f1
f2  A B C D
 a  2 0 2 2
 b  2 2 2 2
 c  2 2 2 2
R> mm <- model.matrix(~ f1 * f2, dd)
R> kappa(mm)                # large, indicating rank deficiency
[1] 4.30923e+16
R> set.seed(1)
R> dd[,"y"] <- mm %*% seq_len(ncol(mm)) +
+                     rnorm(nrow(mm), sd = 0.1)
R> summary(lm(y ~ f1 * f2, dd))   # detects rank deficiency
```

10.2 Motivation: FastLm

```
Call:
lm(formula = y ~ f1 * f2, data = dd)

Residuals:
    Min      1Q  Median      3Q     Max
-0.122  -0.047   0.000   0.047   0.122

Coefficients: (1 not defined because of singularities)
            Estimate Std. Error t value Pr(>|t|)
(Intercept)   0.9779     0.0582    16.8  3.4e-09 ***
f1B          12.0381     0.0823   146.3  < 2e-16 ***
f1C           3.1172     0.0823    37.9  5.2e-13 ***
f1D           4.0685     0.0823    49.5  2.8e-14 ***
f2b           5.0601     0.0823    61.5  2.6e-15 ***
f2c           5.9976     0.0823    72.9  4.0e-16 ***
f1B:f2b      -3.0148     0.1163   -25.9  3.3e-11 ***
f1C:f2b       7.7030     0.1163    66.2  1.2e-15 ***
f1D:f2b       8.9643     0.1163    77.1  < 2e-16 ***
f1B:f2c           NA         NA      NA       NA
f1C:f2c      10.9613     0.1163    94.2  < 2e-16 ***
f1D:f2c      12.0411     0.1163   103.5  < 2e-16 ***
---
Signif. codes:  0 '***' 0.001 '**' 0.01 '*' 0.05 '.' 0.1 ' ' 1

Residual standard error: 0.0823 on 11 degrees of freedom
Multiple R-squared:     1,      Adjusted R-squared:     1
F-statistic: 1.86e+04 on 10 and 11 DF,  p-value: <2e-16

R> summary(fastLm(y ~ f1 * f2, dd)) # some huge coefficients

Call:
fastLm.formula(formula = y ~ f1 * f2, data = dd)

              Estimate       StdErr     t.value   p.value
(Intercept)  2.384e-01    5.091e-01   4.680e-01  0.649605
f1B          5.165e+15    3.394e-01   1.522e+16   < 2e-16 ***
f1C          3.728e+00    7.200e-01   5.177e+00  0.000415 ***
f1D          4.697e+00    7.200e-01   6.523e+00  6.70e-05 ***
f2b          5.752e+00    7.200e-01   7.989e+00  1.19e-05 ***
f2c          6.632e+00    7.200e-01   9.211e+00  3.36e-06 ***
f1B:f2b     -5.165e+15    5.366e-01  -9.625e+15   < 2e-16 ***
f1C:f2b      7.000e+00    1.018e+00   6.875e+00  4.32e-05 ***
f1D:f2b      8.262e+00    1.018e+00   8.114e+00  1.04e-05 ***
f1B:f2c     -5.165e+15    5.366e-01  -9.625e+15   < 2e-16 ***
f1C:f2c      1.027e+01    1.018e+00   1.009e+01  1.46e-06 ***
f1D:f2c      1.131e+01    1.018e+00   1.111e+01  6.02e-07 ***
---
Signif. codes:  0 '***' 0.001 '**' 0.01 '*' 0.05 '.' 0.1 ' ' 1
Multiple R-squared: 0.996,     Adjusted R-squared: 0.993
```

Listing 10.6 An example of a rank-deficient design matrix

The lm() function correctly detected the rank-deficiency in the model matrix which is illustrated by the empty cell in the cross-tabulation. The corresponding interaction parameter has been set to zero; all other coefficients are reasonable. This is achieved by *not* using the standard numerical computing approach, but by relying on a custom pivoting implementation which R provides based on a **Linpack** routine. This will necessarily be slower than the optimized **BLAS** Level 3 routine for QR decomposition called by **Armadillo** for the faster reimplementation

Also, our faster reimplementation has no (explicit) check for rank deficiency by computing κ, the condition number. The model fit is executed in the standard way (as far as numerical computing is concerned), and biased coefficients ensue in this (degenerate) case. Arguably, silently returning wrong results is worse or more inconvenient than an outright failure. This illustrates that it may well pay to bear a small performance penalty by running `lm()` (or at least `lm.fit()`) directly as there may be situations when it is numerically more stable. That said, on current computing architecture floating-point precision tends to be high enough so that such cases are rare in practice. We admit that the example was a little contrived—but it provides a useful illustration.

10.3 Case Study: Kalman Filter Using RcppArmadillo

The **Armadillo** library provides common linear algebra operations in a well-designed and modern C++ framework. It permits us to write elegant and concise code that is also very efficient.

A minor additional focus of **Armadillo** is that it also aims to make it easy for programmers who are familiar with the Matlab / Octave matrix languages to get started in C++ with **Armadillo**. To demonstrate this aspect, we are going to discuss a second example motivated by a discussion of how Matlab can estimate a Kalman filter, and turn the simple program into a C++ version. While R has no comparable tools to convert R code automatically to C or C++, we can, however, achieve rather decent gains by switching the code to **Armadillo** via the **RcppArmadillo** package.

The page `http://www.mathworks.com/products/matlab-coder/demos.html` lists several case studies for this (commercial) code converter. One of these examples covers the Kalman filter. It describes in some detail the original filter, including an example data set, as well as all the steps from the initial script to autogenerated C code.

```
1 %    Copyright 2010 The MathWorks, Inc.
  function y = kalmanfilter(z)
3 %#codegen
  dt=1;
5 % Initialize state transition matrix
  A=[ 1 0 dt 0 0 0;...        % [x  ]
7       0 1 0 dt 0 0;...      % [y  ]
        0 0 1 0 dt 0;...      % [Vx]
9       0 0 0 1 0 dt;...      % [Vy]
        0 0 0 0 1 0 ;...      % [Ax]
11      0 0 0 0 0 1 ];        % [Ay]
```

10.3 Case Study: Kalman Filter Using RcppArmadillo

```matlab
    H = [ 1 0 0 0 0 0; 0 1 0 0 0 0 ];   % Initialize measurement
          matrix
13  Q = eye(6);
    R = 1000 * eye(2);
15  persistent x_est p_est                % Initial state conditions
    if isempty(x_est)
17      x_est = zeros(6, 1);              % x_est=[x,y,Vx,Vy,Ax,Ay]'
        p_est = zeros(6, 6);
19  end
    % Predicted state and covariance
21  x_prd = A * x_est;
    p_prd = A * p_est * A' + Q;
23  % Estimation
    S = H * p_prd' * H' + R;
25  B = H * p_prd';
    klm_gain = (S \ B)';
27  % Estimated state and covariance
    x_est = x_prd + klm_gain * (z - H * x_prd);
29  p_est = p_prd - klm_gain * H * p_prd;
    % Compute the estimated measurements
31  y = H * x_est;
    end                   % of the function
```

Listing 10.7 Basic Kalman filter in Matlab

The R code below reimplements this basic linear Kalman filter.

```r
    FirstKalmanR <- function(pos) {
2
        kalmanfilter <- function(z) {
4           dt <- 1

6           A <- matrix( c( 1, 0, dt, 0, 0, 0,       # x
                            0, 1, 0, dt, 0, 0,       # y
8                           0, 0, 1, 0, dt, 0,       # Vx
                            0, 0, 0, 1, 0, dt,       # Vy
10                          0, 0, 0, 0, 1, 0,        # Ax
                            0, 0, 0, 0, 0, 1),      # Ay
12                          6, 6, byrow=TRUE)
            H <- matrix( c(1, 0, 0, 0, 0, 0,
14                         0, 1, 0, 0, 0, 0),
                            2, 6, byrow=TRUE)
16          Q <- diag(6)
            R <- 1000 * diag(2)
18
            N <- nrow(pos)
20          y <- matrix(NA, N, 2)

22          ## predicted state and covariance
            xprd <- A %*% xest
24          pprd <- A %*% pest %*% t(A) + Q

26          ## estimation
            S <- H %*% t(pprd) %*% t(H) + R
28          B <- H %*% t(pprd)
```

```
        kalmangain <- t(solve(S, B))

        ## est. state and cov., assign to vars in parent env
        xest <<- xprd + kalmangain %*% (z - H %*% xprd)
        pest <<- pprd - kalmangain %*% H %*% pprd

        ## compute the estimated measurements
        y <- H %*% xest
    }

    xest <- matrix(0, 6, 1)
    pest <- matrix(0, 6, 6)

    for (i in 1:N) {
        y[i,] <- kalmanfilter(t(pos[i,,drop=FALSE]))
    }

    invisible(y)
}
```

Listing 10.8 Basic Kalman filter in R

The **Matlab** example uses "persistent" (or "static" for **C++** programmers) variables for xest and pest. We use a different **R** paradigm of defining these variables in the enclosing function. Otherwise, the code is very similar to the original example. Figure 10.1 displays the object trajectory as well as the estimate provided by the Kalman filter.

A slight improvement is available when the invariant code creating variables, from the assignment of dt all the way to the initial variable setup, is also moved to the enclosing function.

```
KalmanR <- function(pos) {

    kalmanfilter <- function(z) {
        ## predicted state and covariance
        xprd <- A %*% xest
        pprd <- A %*% pest %*% t(A) + Q

        ## estimation
        S <- H %*% t(pprd) %*% t(H) + R
        B <- H %*% t(pprd)
        ##  kalmangain <- (S \ B)'
        kalmangain <- t(solve(S, B))

        ## estimated state and covariance, assign to vars in
            parent env
        xest <<- xprd + kalmangain %*% (z - H %*% xprd)
        pest <<- pprd - kalmangain %*% H %*% pprd

        ## compute the estimated measurements
        y <- H %*% xest
    }
```

10.3 Case Study: Kalman Filter Using **RcppArmadillo**

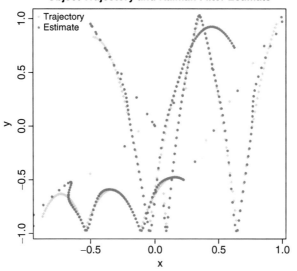

Fig. 10.1 Object trajectory and Kalman filter estimate

```
        dt <- 1
23      A <- matrix( c( 1, 0, dt, 0, 0, 0,      # x
                        0, 1, 0, dt, 0, 0,      # y
25                      0, 0, 1, 0, dt, 0,      # Vx
                        0, 0, 0, 1, 0, dt,      # Vy
27                      0, 0, 0, 0, 1, 0,       # Ax
                        0, 0, 0, 0, 0, 1),      # Ay
29                    6, 6, byrow=TRUE)
        H <- matrix( c(1, 0, 0, 0, 0, 0,
31                     0, 1, 0, 0, 0, 0),
                     2, 6, byrow=TRUE)
33      Q <- diag(6)
        R <- 1000 * diag(2)
35
        N <- nrow(pos)
37      y <- matrix(NA, N, 2)

39      xest <- matrix(0, 6, 1)
        pest <- matrix(0, 6, 6)
41
        for (i in 1:N) {
43          y[i,] <- kalmanfilter(t(pos[i,,drop=FALSE]))
        }
45
        invisible(y)
47  }
```

Listing 10.9 Basic Kalman filter in R

When rewriting this implementation in C++, a first option would be to copy the structure of the first example in Listing 10.8. However, just as Listing 10.9 improves upon the first by factoring out the assignment of variables that are invariant to the call given a particular observation, we can do something very similar in C++ by using basic principles of object-oriented programming. By creating a class for the Kalman filter, we can instantiate the constant variables in the class constructor and be assured that this code will be executed exactly once. The actual estimation—done in the R code in function kalmanfilter—can then be computed in a class member variable.

```
using namespace arma;

class Kalman {
private:
    mat A, H, Q, R, xest, pest;
    double dt;

public:
    // constructor, sets up data structures
    Kalman() : dt(1.0) {
        A.eye(6,6);
        A(0,2) = A(1,3) = A(2,4) = A(3,5) = dt;
        H.zeros(2,6);
        H(0,0) = H(1,1) = 1.0;

        Q.eye(6,6);
        R = 1000 * eye(2,2);

        xest.zeros(6,1);
        pest.zeros(6,6);
    }

    // sole member function: estimate model
    mat estimate(const mat & Z) {

        unsigned int n = Z.n_rows, k = Z.n_cols;
        mat Y = zeros(n, k);
        mat xprd, pprd, S, B, kalmangain;
        colvec z, y;

        for (unsigned int i = 0; i<n; i++) {
            colvec z = Z.row(i).t();

            // predicted state and covariance
            xprd = A * xest;
            pprd = A * pest * A.t() + Q;

            // estimation
            S = H * pprd.t() * H.t() + R;
            B = H * pprd.t();

            // kalmangain = t(S \ B)
```

10.3 Case Study: Kalman Filter Using **RcppArmadillo**

```
                kalmangain = trans(solve(S, B));

                // estimated state and covariance
                xest = xprd + kalmangain * (z - H * xprd);
                pest = pprd - kalmangain * H * pprd;

                // compute the estimated measurements
                y = H * xest;

                Y.row(i) = y.t();
            }
            return Y;
        }
    };
```

Listing 10.10 Basic Kalman filter class in C++ using **Armadillo**

The code itself is very close indeed to the original **Matlab** code. One could argue that it is easier on the eye than the R code. A simple * for matrix or vector multiplication—which we can use as C++ permits one to overload these operators for appropriately defined types such as matrices or vectors—is more succinct than the %*% operator in R.

All the code in Listing 10.10 can be assigned to a variable kalmanClass for included text which we use along with the now very simple function body to create a function via cxxfunction:

```
kalmanSrc <- '
    mat Z = as<mat>(ZS);        // passed from R
    Kalman K;
    mat Y = K.estimate(Z);
    return wrap(Y);
'

KalmanCpp <- cxxfunction(signature(ZS="numeric"),
                         body=kalmanSrc,
                         include=kalmanClass,
                         plugin="RcppArmadillo")
```

Listing 10.11 Basic Kalman filter function in C++

Given the two R versions and this C++ version, we can first determine that results are in fact identical (to numerical precision) between these variants, and then run a benchmark example.

```
R> require(rbenchmark)
R> require(compiler)
R>
R> FirstKalmanRC <- cmpfun(FirstKalmanR)
R> KalmanRC <- cmpfun(KalmanR)
R>
R> stopifnot(identical(KalmanR(pos), KalmanRC(pos)),
+           all.equal(KalmanR(pos), KalmanCpp(pos)),
+           identical(FirstKalmanR(pos), FirstKalmanRC(pos)),
```

```
+              all.equal(KalmanR(pos), FirstKalmanR(pos)))
R>
R> res <- benchmark(KalmanR(pos),
+                   KalmanRC(pos),
+                   FirstKalmanR(pos),
+                   FirstKalmanRC(pos),
+                   KalmanCpp(pos),
+                   columns = c("test", "replications",
+                               "elapsed", "relative"),
+                   order="relative",
+                   replications=100)
R>
R> print(res)
            test replications elapsed relative
5     KalmanCpp(pos)         100   0.087   1.0000
2      KalmanRC(pos)         100   5.774  66.3678
1       KalmanR(pos)         100   6.448  74.1149
4 FirstKalmanRC(pos)         100   8.153  93.7126
3  FirstKalmanR(pos)         100   8.901 102.3103
```

Listing 10.12 Basic Kalman filter timing comparison

The timing results are very satisfactory. Compared to the basic R routine, an improvement in run-time of around 100 times can be achieved. Even the byte-compiled version gains only about ten percent on the basic R versions, and C++ shows an over 90-fold improvement. The simple change of moving invariant code out of the estimation function helps reduce run-time by about a quarter: now the C++ code is about 74 times as fast as the R code, and 66 times as fast as the byte-compiled variant.

10.4 RcppArmadillo and Armadillo Differences

Generally speaking, **RcppArmadillo** does not differ from **Armadillo**. The core source code of the actual **Armadillo** implementation is included "as-is" and not modified.

Armadillo is written to be used as a portable, general-purpose C++ library with the expectation of being used with a variety of compilers and operating systems. In our case, and for the purposes of **RcppArmadillo**, we have a predictable and narrowly defined setup. We know, for example, that there always is an underlying R installation whenever **Rcpp** and **RcppArmadillo** are used.

This enables us to simplify and standardize the use of **Armadillo** by making the following definitions in a configuration header file sourced before the **Armadillo** headers themselves are sources:

```
#define ARMA_USE_LAPACK

#define ARMA_USE_BLAS
```

10.4 RcppArmadillo and Armadillo Differences

```
6 #define ARMA_HAVE_STD_ISFINITE
  #define ARMA_HAVE_STD_ISINF
8 #define ARMA_HAVE_STD_ISNAN
  #define ARMA_HAVE_STD_SNPRINTF
```

Listing 10.13 Standard defines for RcppArmadillo

We can always assume a **Lapack** and **BLAS** installation via R as R will either be built against the system **BLAS** and **Lapack** libraries or provide its own implementation for its usage. Similarly we can make some assumptions about how complete the C library is (though we do undefine all these values on the Solaris platform, and undefine just one for Windows 64).

Two more definitions are more specific to R. Because R provides the "shell" around our statistical computing, programs need to synchronize their (printed) output with R which uses its own buffering. The CRAN maintainers now warn if code uses functions which direct print such as `printf`, or `puts`, or if the C++ facility `std::cout` is used. For standard printing, we can use `Rprintf` from the R API which also provides `REprintf` for error messages. Thanks to a contributed patch, **Rcpp** now wraps a special output device `Rcpp::Rcout` around calls to `Rprintf`. By defining ARMA_DEFAULT_OSTREAM, all output generated by **Armadillo** is then synchronized via the buffering done by R.

```
1 // Rcpp has its own stream object which cooperates more nicely
  // with R's i/o -- and as of Armadillo 2.4.3, we can use this
3 // stream object as well
  #if !defined(ARMA_DEFAULT_OSTREAM)
5 #define ARMA_DEFAULT_OSTREAM Rcpp::Rcout
  #endif
7
  // R now defines NDEBUG which suppresses a number of useful
9 // Armadillo tests  Users can still defined it later, and/or
  // define ARMA_NO_DEBUG
11 #if defined(NDEBUG)
  #undef NDEBUG
13 #endif
```

Listing 10.14 Standard defines for RcppArmadillo

A related matter is the definition of NDEBUG as it (among other things) typically inhibits program halt if a call to `assert` results in a (logically) false condition. This is reasonable from the point of R which can ill afford an exit in a subroutine. However, this has side effects as it may also turn off useful testing. In the case of **Armadillo**, bounds checks for vector and matrix indices are suppressed when NDEBUG is defined. This may not be desirable during the development of new code and is the reason why this definition is removed in the **RcppArmadillo** headers. Users can still define it, or define ARMA_NO_DEBUG if they want it.

Chapter 11
RcppGSL

Abstract The **RcppGSL** package provides an easy-to-use interface between data structures from the GNU Scientific Library, or **GSL** for short, and R by building on facilities provided in the **Rcpp** package. The **GSL** is a well-known collection of numerical routines for scientific computing. It is particularly useful for C and C++ programs as it provides a standard C interface to a wide range of mathematical routines. The chapter provides an introduction to the vector and matrix types in **RcppGSL**, illustrates their use by revisiting the linear modeling example, discusses how to deploy the **RcppGSL** from another package and via **inline**, and closes with an extended application example.

11.1 Introduction

The GNU Scientific Library, or **GSL**, is a collection of routines for scientific computing and numerical analysis (Galassi et al. 2010). It is a rigorously developed and tested library providing support for a wide range of scientific or numerical tasks. Among the topics covered in the **GSL** are complex numbers, roots of polynomials, special functions, vector and matrix data structures, permutations, combinations, sorting, BLAS support, linear algebra, fast Fourier transforms, eigensystems, random numbers, quadrature, random distributions, quasi-random sequences, Monte Carlo integration, N-tuples, differential equations, simulated annealing, numerical differentiation, interpolation, series acceleration, Chebyshev approximations, root-finding, discrete Hankel transforms least-squares fitting, minimization, physical constants, basis splines, and wavelets.

Support for C programming with the **GSL** is readily available. The **GSL** itself is written in C (just like R) and provides a C-language Application Programming Interface (API). Several scripting languages have interfaces to the **GSL** library; the CRAN network for R also contains a package **gsl** providing access to **GSL** functionality for R users.

As a C-language API is provided, access from C++ is also possible, albeit not at the abstraction level that can be offered by dedicated C++ implementations.[1]

The **GSL** is somewhat unique among numerical libraries. Its combination of broad coverage of scientific topics, serious implementation effort, and the use of an Open Source license have led to a fairly wide usage of the library. A number of CRAN packages use the **GSL** directly; and (as of late 2012) nine packages use the CRAN package **gsl** (Hankin 2011) which exposes some parts of the **GSL** to R. This is an indication that the **GSL** is popular among programmers using either the C or C++ language for solving problems in applied science.

At the same time, the **Rcpp** package offers a higher-level abstraction between R and underlying C++ (or C) code. **Rcpp** permits R objects such as vectors, matrices, lists, functions, environments, ..., to be manipulated directly at the C++ level, which alleviates the needs for complicated and error-prone parameter passing and memory allocation. It also permits compact vectorized expressions similar to what can be written in R, but written directly at the C++ level.

The **RcppGSL** package aims to help close the gap between R and the **GSL**. It tries to offer access to **GSL** functions, in particular via the vector and matrix data structures used throughout the **GSL**, while staying closer to the "whole object model" familiar to the R programmer.

The rest of the chapter is organized as follows. The next section shows a motivating example of a fast linear model fit routine using **GSL** functions. The following section discusses support for **GSL** vector types, which is followed by a section on matrices. We close with a case study using B-splines provided by the **GSL** from R.

11.2 Motivation: FastLm

As discussed in Chap. 10, fitting linear models is a key building block of analyzing data and model building. R has a very complete and feature-rich function in `lm()`. It can provide a model fit as well as a number of diagnostic measures, either directly or via the corresponding `summary()` method for linear model fits. The `lm.fit()` function also provides a faster alternative (which is, however, recommended only for advanced users) which provides estimates only and fewer statistics for inference. This sometimes leads to user requests for a routine which is both fast and featureful enough.

The `fastLm` routine shown in Listing 11.1 provides such an implementation (and it preceded the routine based on **RcppArmadillo** from Chap. 10). It uses the **GSL** for the least-squares fitting functions and therefore provides a nice example for **GSL** integration with R, and a direct comparison to the Armadillo-based variant introduced in Chap. 10.

---

[1] Several C++ wrappers for the **GSL** have been written over the years, yet none reached a state of completion comparable to the **GSL** itself. Three such wrapping library are http://cholm.home.cern.ch/cholm/misc/gslmm/, http://gslwrap.sourceforge.net/, and http://code.google.com/p/gslcpp/.

11.2 Motivation: FastLm

```
#include <RcppGSL.h>
#include <gsl/gsl_multifit.h>
#include <cmath>

extern "C" SEXP fastLm(SEXP ys, SEXP Xs) {

  try {
    RcppGSL::vector<double> y = ys;       // gsl data str. via SEXP
    RcppGSL::matrix<double> X = Xs;

    int n = X.nrow(), k = X.ncol();
    double chisq;

    RcppGSL::vector<double> coef(k);      // hold the coef vector
    RcppGSL::matrix<double> cov(k,k);     // and covariance matrix

    // the actual fit req. working memory we allocate and free
    gsl_multifit_linear_workspace *work =
                                gsl_multifit_linear_alloc (n, k);
    gsl_multifit_linear (X, y, coef, cov, &chisq, work);
    gsl_multifit_linear_free (work);

    // extract the diagonal as a vector view
    gsl_vector_view diag = gsl_matrix_diagonal(cov) ;

    // currently no direct interface in Rcpp::NumericVector
    // that uses wrap(), so we have to do it in two steps
    Rcpp::NumericVector std_err ; std_err = diag;
    std::transform(std_err.begin(), std_err.end(),
                   std_err.begin(), sqrt);

    Rcpp::List res =
       Rcpp::List::create(Rcpp::Named("coefficients") = coef,
                          Rcpp::Named("stderr") = std_err,
                          Rcpp::Named("df") = n - k);

    // free all the GSL vectors and matrices -- as these are
    // really C data structures we cannot take advantage of
    // automatic C++ memory management
    coef.free(); cov.free(); y.free(); X.free();

    return res;            // return the result list to R

  } catch( std::exception &ex ) {
    forward_exception_to_r( ex );
  } catch(...) {
    ::Rf_error( "c++ exception (unknown reason)" );
  }
  return R_NilValue; // -Wall
}
```

Listing 11.1 FastLm function using **RcppGSL**

We first initialize a **RcppGSL** vector and matrix, each templated to the standard numeric type `double` (and the **GSL** supports other types ranging from lower precision floating point to signed and unsigned integers as well as complex numbers). Next, we reserve another vector and matrix to hold the resulting coefficient estimates as well as the estimate of the covariance matrix. We then allocate workspace using a **GSL** routine, fit the linear model, and free the workspace. This is followed by extraction of the diagonal element from the covariance matrix. We then employ a so-called iterator—a common C++ idiom from the Standard Template Library (STL)—to iterate over the vector of diagonal and transforming it by applying the square root function to compute our standard error of the estimate. Finally, we create a named list with the return value before we free temporary memory allocation. This last step is required because the underlying objects are really C objects conforming to the **GSL** interface. Hence, they do have the automatic memory management we could have with C++ vector or matrix structures as used through the **Rcpp** package. Finally, we return the result to R.

As seen in the previous chapter, **RcppArmadillo** (François et al. 2012) implements a matching `fastLm` function using the Armadillo library by Sanderson (2010) and can do so with more compact code due to C++ features.

11.3 Vectors

This section details the different vector representations, starting with their definition inside the **GSL** itself. We then discuss our layering before showing how the two types map to each other. A discussion of read-only "vector view" classes concludes the section.

11.3.1 GSL Vectors

GSL defines various vector types to manipulate one-dimensional data, similar to R arrays. For example, the `gsl_vector` and `gsl_vector_int` structs are defined as:

```
typedef struct{
   size_t size;
   size_t stride;
   double *data;
   gsl_block *block;
   int owner;
} gsl_vector;

typedef struct {
   size_t size;
   size_t stride;
   int *data;
```

11.3 Vectors

```
    gsl_block_int *block;
14  int owner;
  } gsl_vector_int;
```

Listing 11.2 Definition of gsl_vector and gsl_vector_int

A typical use of the `gsl_vector` struct is given below:

```
1  int i;
   // allocate a gsl_vector of size 3
3  gsl_vector * v = gsl_vector_alloc (3);

5  // fill the vector
   for (i = 0; i < 3; i++) {
7    gsl_vector_set (v, i, 1.23 + i);
   }
9
   // access elements
11 double sum = 0.0 ;
   for (i = 0; i < 3; i++) {
13   sum += gsl_vector_get( v, i ) ;
   }
15
   // free the memory
17 gsl_vector_free (v);
```

Listing 11.3 Example use of gsl_vector

11.3.2 RcppGSL::vector

RcppGSL defines the template `RcppGSL::vector<T>`. It can manipulate pointers to `gsl_vector` objects by taking advantage of C++ templates. With this new type, the previous example becomes:

```
1  int i;
   // allocate a gsl_vector of size 3
3  RcppGSL::vector<double> v(3);

5  // fill the vector
   for (i = 0; i < 3; i++) {
7    v[i] = 1.23 + i ;
   }
9
   // access elements
11 double sum = 0.0 ;
   for (i = 0; i < 3; i++) {
13   sum += v[i] ;
   }
15
   // free the memory
17 v.free() ;
```

Listing 11.4 Example use of RcppGSL::vector<T>

The class RcppGSL::vector<double> implements a smart pointer which can be used anywhere in place of a raw pointer to gsl_vector. Examples are the gsl_vector_set and gsl_vector_get functions above.

Beyond the convenience of a nicer syntax for allocation and release of memory, the RcppGSL::vector template facilitates the interchange of **GSL** vectors with **Rcpp** objects, and hence **R** objects. The following example defines a .Call() compatible function called sum_gsl_vector_int that operates on a gsl_vector_int through the RcppGSL::vector<int> template specialization:

```
RCPP_FUNCTION_1(int, sum_gsl_vector_int,
                RcppGSL::vector<int> vec) {
  int res = std::accumulate(vec.begin(), vec.end(), 0);
  vec.free();  // we need to free vec after use
  return res;
}
```

Listing 11.5 Example RcppGSL::vector<T> function

The macro RCPP_FUNCTION_1 expands its arguments to a single-parameter function. The generated function returns the type given as the first macro argument, has the function name provided by the second macro argument, and takes the third macro argument as the function argument.

Hence, the function can then be called from R as:

```
R> .Call( "sum_gsl_vector_int", 1:10 )
[1] 55
```

Listing 11.6 Example call of RcppGSL::vector<T> function

A second example shows a simple function that grabs elements of an R list as gsl_vector objects using implicit conversion mechanisms of **Rcpp**

```
RCPP_FUNCTION_1(double, gsl_vector_sum_2,
                Rcpp::List data ) {
  // grab "x" as a gsl_vector through
  // the RcppGSL::vector<double> class
  RcppGSL::vector<double> x = data["x"] ;

  // grab "y" as a gsl_vector through
  // the RcppGSL::vector<int> class
  RcppGSL::vector<int> y = data["y"] ;
  double res = 0.0 ;
  for( size_t i=0; i< x->size; i++){
    res += x[i] * y[i] ;
  }

  // we need to explicitly free the memory
  x.free() ;
```

11.3 Vectors

```
     y.free() ;
18
     // return the result
20   return res ;
   }
```

Listing 11.7 Second example RcppGSL::vector<T> function

which can be called from R as follows:

```
1  R> .Call( "gsl_vector_sum_2", data )
   [1] 36.66667
```

Listing 11.8 Example call of second RcppGSL::vector<T> function

11.3.3 Mapping

Table 11.1 shows the mapping between types defined by the **GSL** and their corresponding types in the **RcppGSL** package.

Table 11.1 Correspondence between **GSL** vector types and templates defined in **RcppGSL**

gsl vector	RcppGSL (with RcppGSL:: prefix)
gsl_vector	vector<double>
gsl_vector_int	vector<int>
gsl_vector_float	vector<float>
gsl_vector_long	vector<long>
gsl_vector_char	vector<char>
gsl_vector_complex	vector<gsl_complex>
gsl_vector_complex_float	vector<gsl_complex_float>
gsl_vector_complex_long_double	vector<gsl_complex_long_double>
gsl_vector_long_double	vector<long double>
gsl_vector_short	vector<short>
gsl_vector_uchar	vector<unsigned char>
gsl_vector_uint	vector<unsigned int>
gsl_vector_ushort	vector<insigned short>
gsl_vector_ulong	vector<unsigned long>

11.3.4 Vector Views

Several **GSL** algorithms return **GSL** vector views as their result type. **RcppGSL** defines the template class RcppGSL::vector_view to handle vector views using C++ syntax.

```
   extern "C" SEXP test_gsl_vector_view(){
2    int n = 10 ;
     RcppGSL::vector<double> v(n) ;
4    for( int i=0 ; i<n; i++){
       v[i] = i ;
6    }
     RcppGSL::vector_view<double> v_even =
8        gsl_vector_subvector_with_stride(v,0,2,n/2);
     RcppGSL::vector_view<double> v_odd  =
10       gsl_vector_subvector_with_stride(v,1,2,n/2);

12   List res = List::create(
       _["even"] = v_even,
14     _["odd" ] = v_odd
       ) ;
16   v.free() ; // we only free v, views do not own data
     return res ;
18 }
```

Listing 11.9 Example of a vector view class

As with vectors, C++ objects of type RcppGSL::vector_view can be converted implicitly to their associated **GSL** view type. Table 11.2 displays the pairwise correspondence so that the C++ objects can be passed to compatible **GSL** algorithms. Note that the vector_view<gsl_complex_long_double> variant has been omitted for typesetting reasons.

Table 11.2 Correspondence between **GSL** vector view types and templates defined in **RcppGSL**

gsl vector views	RcppGSL (with RcppGSL:: prefix)
gsl_vector_view	vector_view<double>
gsl_vector_view_int	vector_view<int>
gsl_vector_view_float	vector_view<float>
gsl_vector_view_long	vector_view<long>
gsl_vector_view_char	vector_view<char>
gsl_vector_view_complex	vector_view<gsl_complex>
gsl_vector_view_complex_float	vector_view<gsl_complex_float>
gsl_vector_view_long_double	vector_view<long double>
gsl_vector_view_short	vector_view<short>
gsl_vector_view_uchar	vector_view<unsigned char>
gsl_vector_view_uint	vector_view<unsigned int>
gsl_vector_view_ushort	vector_view<insigned short>
gsl_vector_view_ulong	vector_view<unsigned long>

The vector view class also contains a conversion operator to automatically transform the data of the view object to a **GSL** vector object. This enables the use of vector views where **GSL** would expect a vector.

11.4 Matrices

The **GSL** also defines a set of matrix data types: `gsl_matrix`, `gsl_matrix_int` etc., for which **RcppGSL** defines a corresponding convenience C++ wrapper generated by the `RcppGSL::matrix` template.

11.4.1 Creating Matrices

The `RcppGSL::matrix` template exposes three constructors.

```
  // convert an R matrix to a GSL matrix
2 matrix( SEXP x) throw(::Rcpp::not_compatible)

4 // encapsulate a GSL matrix pointer
  matrix( gsl_matrix* x)
6
  // create a new matrix with the
8 // given number of rows and columns
  matrix( int nrow, int ncol)
```

Listing 11.10 Example use **RcppGSL** matrix class

11.4.2 Implicit Conversion

`RcppGSL::matrix` defines implicit conversion of a pointer to the associated **GSL** matrix type, as well as dereferencing operators, making the class `RcppGSL::matrix` look and feel like a pointer to a **GSL** matrix type.

```
1 gsltype* data ;
  operator gsltype*(){ return data ; }
3 gsltype* operator->() const { return data; }
  gsltype& operator*() const { return *data; }
```

Listing 11.11 Implicit conversion for **RcppGSL** matrix class

11.4.3 Indexing

Indexing of elements of **GSL** matrices is usually done using the *getter* functions `gsl_matrix_get`, `gsl_matrix_int_get`, etc. and the *setter* functions `gsl_matrix_set`, `gsl_matrix_int_set`, etc. As C functions, these have to supply the row and column indices individually.

RcppGSL takes advantage of both operator overloading and templates to make indexing a **GSL** matrix much more convenient and closer to our common mathematical notation as shown in the next example.

```
  // create a matrix of size 10x10
2 RcppGSL::matrix<int> mat(10,10);

4 // fill the diagonal
  for( int i=0; i<10: i++) {
6     mat(i,i) = i ;
  }
```

Listing 11.12 Indexing for **RcppGSL** matrix class

11.4.4 Methods

The RcppGSL::matrix type also defines the following member functions:

nrow() to extract the number of rows
ncol() to extract the number of columns
size() to extract the number of elements
free() to release the memory

11.4.5 Matrix Views

Similar to the vector views discussed above, the **RcppGSL** also provides an implicit conversion operator which returns the underlying matrix stored in the matrix view class.

11.5 Using RcppGSL in Your Package

The **RcppGSL** package contains a complete example providing a single function colNorm which computes a norm for each column of a supplied matrix. This example adapts a matrix example from the **GSL** manual that has been chosen merely as a means for showing how to set up a package to use **RcppGSL**.

Needless to say, we could compute such a matrix norm easily in R using existing facilities. One such possibility is a simple expression as in Listing 11.13

11.5 Using **RcppGSL** in Your Package

which is also shown on the corresponding help page in the example package inside **RcppGSL**.

```
1    apply(M, 2, function(x) sqrt(sum(x^2)))
```
Listing 11.13 Matrix norm in R

One point in favor of using the **GSL** code is that it employs BLAS functions. On sufficiently large matrices, and with suitable BLAS libraries installed, this variant could be faster due to the optimized code in high-performance BLAS libraries and/or the inherent parallelism a multi-core BLAS variant which compute the vector norm in parallel. On all "reasonable" matrix sizes, however, the performance differences should be negligible.

11.5.1 The `configure` Script

11.5.1.1 Using autoconf

Using **RcppGSL** means employing both the **GSL** and R. We may need to find the location of the **GSL** headers and library, and this can be done easily from a `configure` source script which **autoconf** generates from a `configure.in` source file such as the following:

```
1  AC_INIT([RcppGSLExample], 0.1.0)

3  ## Use gsl-config to find arguments for compiler + linker flags
   ##
5  ## Check for non-standard programs: gsl-config(1)
   AC_PATH_PROG([GSL_CONFIG], [gsl-config])
7  ## If gsl-config was found, let's use it
   if test "${GSL_CONFIG}" != ""; then
9      # Use gsl-config for header and linker arguments
       # (without BLAS which we get from R)
11     GSL_CFLAGS=`${GSL_CONFIG} --cflags`
       GSL_LIBS=`${GSL_CONFIG} --libs-without-cblas`
13 else
       AC_MSG_ERROR([gsl-config not found, is GSL installed?])
15 fi

17 ## Use Rscript to query Rcpp for compiler and linker flags
   ## link flag providing library as well as path to library,
19 ## and optionally rpath
   RCPP_LDFLAGS=`${R_HOME}/bin/Rscript -e 'Rcpp:::LdFlags()'`
21
   # Now substitute these variables
23 # in src/Makevars.in to create src/Makevars
   AC_SUBST(GSL_CFLAGS)
25 AC_SUBST(GSL_LIBS)
```

```
                AC_SUBST(RCPP_LDFLAGS)
27
                AC_OUTPUT(src/Makevars)
```

Listing 11.14 Autoconf script for **RcppGSL** use

Such a source `configure.in` gets converted into a script `configure` by invoking the `autoconf` program.

11.5.1.2 Using Functions Provided by RcppGSL

RcppGSL provides R functions that allow one to retrieve the same information. Therefore, the configure script can also be written as:

```
  #!/bin/sh
2
  GSL_CFLAGS=`${R_HOME}/bin/Rscript -e "RcppGSL:::CFlags()"`
4 GSL_LIBS=`${R_HOME}/bin/Rscript -e "RcppGSL:::LdFlags()"`
  RCPP_LDFLAGS=`${R_HOME}/bin/Rscript -e "Rcpp:::LdFlags()"`
6
   sed -e "s|@GSL_LIBS@|${GSL_LIBS}|" \
8      -e "s|@GSL_CFLAGS@|${GSL_CFLAGS}|" \
       -e "s|@RCPP_LDFLAGS@|${RCPP_LDFLAGS}|" \
10     src/Makevars.in > src/Makevars
```

Listing 11.15 Shell script configuration script for **RcppGSL** use

Similarly, the configure.win for windows can be written as:

```
  RSCRIPT="${R_HOME}/bin${R_ARCH_BIN}/Rscript.exe"
2 GSL_CFLAGS=`${RSCRIPT} -e "RcppGSL:::CFlags()"`
  GSL_LIBS=`${RSCRIPT} -e "RcppGSL:::LdFlags()"`
4 RCPP_LDFLAGS=`${RSCRIPT} -e "Rcpp:::LdFlags()"`
6 sed -e "s|@GSL_LIBS@|${GSL_LIBS}|" \
      -e "s|@GSL_CFLAGS@|${GSL_CFLAGS}|" \
8     -e "s|@RCPP_LDFLAGS@|${RCPP_LDFLAGS}|" \
      src/Makevars.in > src/Makevars.win
```

Listing 11.16 Windows shell script configuration script for **RcppGSL** use

11.5.2 The `src` Directory

The C++ source file takes the matrix supplied from R and applies the GSL function to each column.

```
1
  #include <RcppGSL.h>
3 #include <gsl/gsl_matrix.h>
```

11.5 Using **RcppGSL** in Your Package

```
#include <gsl/gsl_blas.h>

extern "C" SEXP colNorm(SEXP sM) {

  try {
    // create gsl data structures from SEXP
    RcppGSL::matrix<double> M = sM;
    int k = M.ncol();
    Rcpp::NumericVector n(k);          // for results

    for (int j = 0; j < k; j++) {
      RcppGSL::vector_view<double> colview =
            gsl_matrix_column (M, j);
      n[j] = gsl_blas_dnrm2(colview);
    }
    M.free() ;
    return n;                          // return vector

  } catch( std::exception &ex ) {
    forward_exception_to_r( ex );

  } catch(...) {
    ::Rf_error( "c++ exception (unknown reason)" );
  }
  return R_NilValue; // -Wall
}
```

Listing 11.17 Vector norm function for **RcppGSL**

The `Makevars.in` file governs the compilation and uses the values supplied by `configure` during build-time:

```
# set by configure
GSL_CFLAGS = @GSL_CFLAGS@
GSL_LIBS   = @GSL_LIBS@
RCPP_LDFLAGS = @RCPP_LDFLAGS@

# combine with standard arguments for R
PKG_CPPFLAGS = $(GSL_CFLAGS)
PKG_LIBS = $(GSL_LIBS) $(RCPP_LDFLAGS)
```

Listing 11.18 `Makevars.in` for **RcppGSL** example

The variables surrounded by @ will be filled by `configure` during package build-time with values determined by the `configure` code shown above.

11.5.3 The R Directory

The R source is very simple: a single matrix is passed to C++:

```
1 colNorm <- function(M) {
      stopifnot(is.matrix(M))
3     res <- .Call("colNorm", M, PACKAGE="RcppGSLExample")
  }
```

Listing 11.19 R function for **RcppGSL** example

11.6 Using RcppGSL with inline

As we have seen throughout the book, the **inline** package (Sklyar et al. 2012) is very helpful for prototyping code in C, C++, or Fortran as it takes care of code compilation, linking and dynamic loading directly from R. It is being used extensively by **Rcpp**, for example, in the numerous unit tests.

The example below shows how **inline** can be deployed with **RcppGSL**. We implement the same column norm example, but this time as an R script which is compiled, linked, and loaded on-the-fly. Compared to standard use of **inline**, we have to make sure to add a short section declaring which header files from **GSL** we need to use; the **RcppGSL** then communicates with **inline** to tell it about the location and names of libraries used to build code against **GSL**.

```
  R> require(inline)
2 R> inctxt='
  +     #include <gsl/gsl_matrix.h>
4 +     #include <gsl/gsl_blas.h>
  +   '
6 R> bodytxt='
  +     // create gsl data structures from SEXP
8 +     RcppGSL::matrix<double> M = sM;
  +     int k = M.ncol();
10 +    Rcpp::NumericVector n(k);       // for results
  +
12 +    for (int j = 0; j < k; j++) {
  +        RcppGSL::vector_view<double> colview =
14 +            gsl_matrix_column (M, j);
  +        n[j] = gsl_blas_dnrm2(colview);
16 +    }
  +    M.free() ;
18 +    return n;                       // return vector
  +   '
20 R> foo <- cxxfunction(
  +        signature(sM="numeric"),
22 +       body=bodytxt, inc=inctxt, plugin="RcppGSL")
  R> ## see Section 8.4.13 of the GSL manual:
24 R> ## create M as a sum of two outer products
  R> M <- outer(sin(0:9), rep(1,10), "*") +
26 +        outer(rep(1, 10), cos(0:9), "*")
  R> foo(M)
```

```
28 [1] 4.314614  3.120504  2.193159  3.261141  2.534157
   [6] 2.572810  4.204689  3.652017  2.085236  3.073134
```

Listing 11.20 Using **RcppGSL** with **inline**

The RcppGSL inline plugin supports creation of a package skeleton based on the inline function.

```
1 R> package.skeleton( "mypackage", foo )
```

Listing 11.21 Using package.skeleton with **inline** result

This creates a skeleton package, similar to what we have seen in Chap. 5, based on the function produced by cxxfunction() and assigned to the function object foo() as per the code example in Listing 11.20.

11.7 Case Study: GSL-Based B-Spline Fit Using RcppGSL

The **GSL** and **R** both overlap in a number of areas. In that sense the following example is contrived as we could compute it entirely in **R**. However, as the **GSL** is a well-established numerical library in its own right, it is of interest to show how examples from **GSL** can be deployed together with **R**.

In this section, we illustrate this using an example from Section 39.7 of the **GSL** reference manual. We generate data from

$$y(x) = e^{-x/10} \cos(x) \qquad \text{with} \quad x \in [0, 15]$$

and fit this data using weighted least squares using a cubic B-spline basis function with uniform breakpoints. Again, we are not interested as much in the statistical aspect of this problem as we are in exploring how to let **R** use code from the **GSL**.

The original program follows in Listing 11.22.

```
1 #include <stdio.h>
  #include <stdlib.h>
3 #include <math.h>
  #include <gsl/gsl_bspline.h>
5 #include <gsl/gsl_multifit.h>
  #include <gsl/gsl_rng.h>
7 #include <gsl/gsl_randist.h>
  #include <gsl/gsl_statistics.h>
9
  /* number of data points to fit */
11 #define N        200

13 /* number of fit coefficients */
  #define NCOEFFS  12
15
  /* nbreak = ncoeffs + 2 - k = ncoeffs - 2 since k = 4 */
```

```
17 #define NBREAK   (NCOEFFS - 2)

19 int main (void) {
       const size_t n = N;
21     const size_t ncoeffs = NCOEFFS;
       const size_t nbreak = NBREAK;
23     size_t i, j;
       gsl_bspline_workspace *bw;
25     gsl_vector *B;
       double dy;
27     gsl_rng *r;
       gsl_vector *c, *w;
29     gsl_vector *x, *y;
       gsl_matrix *X, *cov;
31     gsl_multifit_linear_workspace *mw;
       double chisq, Rsq, dof, tss;
33
       gsl_rng_env_setup();
35     r = gsl_rng_alloc(gsl_rng_default);

37     /* allocate a cubic bspline workspace (k = 4) */
       bw = gsl_bspline_alloc(4, nbreak);
39     B = gsl_vector_alloc(ncoeffs);

41     x = gsl_vector_alloc(n);
       y = gsl_vector_alloc(n);
43     X = gsl_matrix_alloc(n, ncoeffs);
       c = gsl_vector_alloc(ncoeffs);
45     w = gsl_vector_alloc(n);
       cov = gsl_matrix_alloc(ncoeffs, ncoeffs);
47     mw = gsl_multifit_linear_alloc(n, ncoeffs);

49     printf("#m=0,S=0\n");
       /* this is the data to be fitted */
51
       for (i = 0; i < n; ++i) {
53         double sigma;
           double xi = (15.0 / (N - 1)) * i;
55         double yi = cos(xi) * exp(-0.1 * xi);

57         sigma = 0.1 * yi;
           dy = gsl_ran_gaussian(r, sigma);
59         yi += dy;

61         gsl_vector_set(x, i, xi);
           gsl_vector_set(y, i, yi);
63         gsl_vector_set(w, i, 1.0 / (sigma * sigma));

65         printf("%f %f\n", xi, yi);
       }
67
       /* use uniform breakpoints on [0, 15] */
69     gsl_bspline_knots_uniform(0.0, 15.0, bw);
```

11.7 Case Study: GSL-Based B-Spline Fit Using **RcppGSL**

```
        /* construct the fit matrix X */
        for (i = 0; i < n; ++i) {
            double xi = gsl_vector_get(x, i);

            /* compute B_j(xi) for all j */
            gsl_bspline_eval(xi, B, bw);

            /* fill in row i of X */
            for (j = 0; j < ncoeffs; ++j) {
                double Bj = gsl_vector_get(B, j);
                gsl_matrix_set(X, i, j, Bj);
            }
        }

        /* do the fit */
        gsl_multifit_wlinear(X, w, y, c, cov, &chisq, mw);

        dof = n - ncoeffs;
        tss = gsl_stats_wtss(w->data, 1, y->data, 1, y->size);
        Rsq = 1.0 - chisq / tss;

        fprintf(stderr, "chisq/dof = %e, Rsq = %f\n", chisq / dof,
            Rsq);

        /* output the smoothed curve */
        {
            double xi, yi, yerr;

            printf("#m=1,S=0\n");
            for (xi = 0.0; xi < 15.0; xi += 0.1) {
                gsl_bspline_eval(xi, B, bw);
                gsl_multifit_linear_est(B, c, cov, &yi, &yerr);
                printf("%f %f\n", xi, yi);
            }
        }

        gsl_rng_free(r);
        gsl_bspline_free(bw);
        gsl_vector_free(B);
        gsl_vector_free(x);
        gsl_vector_free(y);
        gsl_matrix_free(X);
        gsl_vector_free(c);
        gsl_vector_free(w);
        gsl_matrix_free(cov);
        gsl_multifit_linear_free(mw);

        return 0;
    } /* main() */
```

Listing 11.22 B-spline fit example from the **GSL**

This original GSL example provides a stand-alone program with a single `main()` function. First, the data is generated and written out to the standard output. Next, the cubic B-spline is set up and fit, and the result is written out. In the original documentation it is then suggested to use an external plotting program to visualize data and fit. We can of course easily do the second step in R by reading the input, subsetting it into input data (of which there are 200 lines) and results data (151 lines for the grid from 0.0 to 15.0 in increments of 0.1.

In order to use this functionality from R, we will decompose the program into two parts: data generation and data fitting. Each part will be addressed by a single C++ function which executes just its part.

We use "Rcpp attributes" (described in Sect. 2.6) to provide access to this C++ code from R itself.

```
// [[Rcpp::depends(RcppGSL)]]
#include <RcppGSL.h>

#include <gsl/gsl_bspline.h>
#include <gsl/gsl_multifit.h>
#include <gsl/gsl_rng.h>
#include <gsl/gsl_randist.h>
#include <gsl/gsl_statistics.h>

const int N = 200;                     // number of data points to fit
const int NCOEFFS = 12;                // number of fit coefficients
const int NBREAK = (NCOEFFS - 2);      // nbreak=ncoeffs-2 since k = 4
```

Listing 11.23 Beginning of C++ file with B-spline fit for R

This first declares a dependency on **RcppGSL** implying that R will use both the header files and library from the **GSL**—by using the plugin discussed above. Several header files are then included to declare the types used by **RcppGSL** (and **Rcpp**) as well as for the **GSL** functionality needed.

We can then define the first function to generate the data.

```
// [[Rcpp::export]]
Rcpp::List genData() {

    const size_t n = N;
    size_t i;
    double dy;
    gsl_rng *r;
    RcppGSL::vector<double> w(n), x(n), y(n);

    gsl_rng_env_setup();
    r = gsl_rng_alloc(gsl_rng_default);

    //printf("#m=0,S=0\n");
    /* this is the data to be fitted */

    for (i = 0; i < n; ++i) {
        double sigma;
        double xi = (15.0 / (N - 1)) * i;
        double yi = cos(xi) * exp(-0.1 * xi);

```

11.7 Case Study: GSL-Based B-Spline Fit Using **RcppGSL**

```
            sigma = 0.1 * yi;
22          dy = gsl_ran_gaussian(r, sigma);
            yi += dy;
24
            gsl_vector_set(x, i, xi);
26          gsl_vector_set(y, i, yi);
            gsl_vector_set(w, i, 1.0 / (sigma * sigma));
28
            //printf("%f %f\n", xi, yi);
30      }

32      Rcpp::DataFrame res =
            Rcpp::DataFrame::create(Rcpp::Named("x") = x,
34                                  Rcpp::Named("y") = y,
                                    Rcpp::Named("w") = w);
36
        x.free();
38      y.free();
        w.free();
40      gsl_rng_free(r);

42      return(res);
    }
```

Listing 11.24 Data generation for **GSL** B-spline fit for R

Similarly, the second function used to fit the data can be defined as well.

```
1   // [[Rcpp::export]]
    Rcpp::List fitData(Rcpp::DataFrame ds) {
3
        const size_t ncoeffs = NCOEFFS;
5       const size_t nbreak = NBREAK;

7       const size_t n = N;
        size_t i, j;
9
        Rcpp::DataFrame D(ds);                   // construct data.frame
11      RcppGSL::vector<double> y = D["y"]; // access columns by name
        RcppGSL::vector<double> x = D["x"]; // assigning GSL vectors
13      RcppGSL::vector<double> w = D["w"];

15      gsl_bspline_workspace *bw;
        gsl_vector *B;
17      gsl_vector *c;
        gsl_matrix *X, *cov;
19      gsl_multifit_linear_workspace *mw;
        double chisq, Rsq, dof, tss;
21
        // allocate a cubic bspline workspace (k = 4)
23      bw = gsl_bspline_alloc(4, nbreak);
        B = gsl_vector_alloc(ncoeffs);
25
        X = gsl_matrix_alloc(n, ncoeffs);
27      c = gsl_vector_alloc(ncoeffs);
```

```
        cov = gsl_matrix_alloc(ncoeffs, ncoeffs);
        mw = gsl_multifit_linear_alloc(n, ncoeffs);

        // use uniform breakpoints on [0, 15]
        gsl_bspline_knots_uniform(0.0, 15.0, bw);

        // construct the fit matrix X
        for (i = 0; i < n; ++i) {
            double xi = gsl_vector_get(x, i);

            // compute B_j(xi) for all j
            gsl_bspline_eval(xi, B, bw);

            // fill in row i of X
            for (j = 0; j < ncoeffs; ++j) {
                double Bj = gsl_vector_get(B, j);
                gsl_matrix_set(X, i, j, Bj);
            }
        }

        // do the fit
        gsl_multifit_wlinear(X, w, y, c, cov, &chisq, mw);

        dof = n - ncoeffs;
        tss = gsl_stats_wtss(w->data, 1, y->data, 1, y->size);
        Rsq = 1.0 - chisq / tss;

        // output the smoothed curve
        Rcpp::NumericMatrix M(150,2);
        double xi, yi, yerr;
        for (xi = 0.0, i=0; xi < 15.0; xi += 0.1, i++) {
            gsl_bspline_eval(xi, B, bw);
            gsl_multifit_linear_est(B, c, cov, &yi, &yerr);
            M(i,0) = xi;
            M(i,1) = yi;
        }

        gsl_bspline_free(bw);
        gsl_vector_free(B);
        gsl_matrix_free(X);
        gsl_vector_free(c);
        gsl_matrix_free(cov);
        gsl_multifit_linear_free(mw);

        return(Rcpp::List::create(
                Rcpp::Named("M")=M,
                Rcpp::Named("chisqdof")=Rcpp::wrap(chisq/dof),
                Rcpp::Named("rsq")=Rcpp::wrap(Rsq)));
}
```

Listing 11.25 Data fit for **GSL** B-spline with R

Finally, we can generate the compiled functions, generate the data, and fit the spline model. This fit is illustrated in a chart shown in Fig. 11.1.

11.7 Case Study: GSL-Based B-Spline Fit Using RcppGSL

```
2  # compile two functions
   sourceCpp("bSpline.cpp")
4
   # generate the data
6  dat <- genData()

8  # fit the model, returns matrix and gof measures
   fit <- fitData(dat)
10 M <- fit[[1]]

12 # plot
   op <- par(mar=c(3,3,1,1))
14 plot(dat[,"x"], dat[,"y"], pch=19, col="#00000044")
   lines(M[,1], M[,2], col="orange", lwd=2)
16 par(op)
```

Listing 11.26 R side of GSL B-spline example

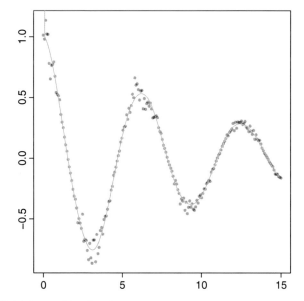

Fig. 11.1 Artificial data and B-spline fit

Chapter 12
RcppEigen

Abstract The **RcppEigen** package provides an interface to the **Eigen** library. **Eigen** is a featureful C++ library which deploys modern template meta-programming techniques. It is similar to **Armadillo** but provides an even more granular application-programming interface (API). This chapter provides an introduction to the **Rcpp Eigen** package by introducing the core data structures, illustrating some of the available matrix decomposition methods and concludes with a case study of particular C++ implementation (providing what is called a "factory" pattern) for different matrix decomposition approaches in order to provide a faster reimplementation of the `lm` method.

12.1 Introduction

Eigen is a modern C++ library for linear algebra, similar in scope as **Armadillo** (which was discussed in Chap. 10), but with an even more fine-grained application-programming interface (API). **Eigen** (Guennebaud et al. 2012) started as a sub-project to KDE (a popular Linux desktop environment), initially focusing on fixed-size matrices to represent rotations, projections, or affine transformations in a visualization application. **Eigen** grew from there and has over the course of about a decade produced three major releases with "Eigen3" being the current major version. **Eigen** is now widely used in a number of projects, including **ceres**, a large-scale nonlinear least-squares solver released by Google.[1]

And just like **Armadillo**, **Eigen** has been prepared for use with **Rcpp** by providing appropriate conversion functions `as<>()` and `wrap()` in the **RcppEigen** package (Bates and Eddelbuettel 2013). The next section introduces some of the key data types in **Eigen**, as well as the corresponding converters accessible from **Rcpp**.

---

[1] See https://code.google.com/p/ceres-solver/.

12.2 Eigen Classes

12.2.1 Fixed-Size Vectors and Matrices

The earliest version of **Eigen** aimed at supporting visualizations and projections in computational chemistry. For this task, fixed-size matrices and vectors are appropriate and are still supported in the current version.

C++ meta-template programming (Abrahams and Gurtovoy 2004) is used extensively throughout **Eigen**. When dimensions are known at compile-time, operations which would commonly involve loops at run-time can in fact be converted at compile-time. Consider this simple example:

```
  Eigen::Vector3d x(1,2,3);
2 Eigen::Vector3d y(4,5,6);

4 Eigen::Matrix3d m1 = x * y.transpose();
         double m2 = x.transpose() * y;
6
  Rcpp::Rcout << "Outer:\n" << m1 << std::endl;
8 Rcpp::Rcout << "Inner:\n" << m2 << std::endl;
```

Listing 12.1 A simple **Eigen** example using fixed-size vectors and matrices

The length of the two vectors is fixed at size three in the definition of the vector classes. The (square) matrix is also of size three.

The prime reason for creating the fixed-size variants is efficiency. Through the use of templates, this library lets the compiler create a more efficient implementation for the inner and outer products which, essentially, replaces the loop constructs with a constant assignment. We will show just how dramatic the difference can be below.

Conceptually, the representation in **Eigen** is of the following type (where we limit ourselves to dimension three, but variants for dimension two and four exist, as do variants for types `float` and `complex` not shown here):

```
  typedef Matrix<int,    3, 1>   Vector3i;
2 typedef Matrix<double, 3, 1>   Vector3d;

4 typedef Matrix<int,    1, 3>   RowVector3i;
  typedef Matrix<int,    3, 1>   ColVector3i;
6 typedef Matrix<double, 1, 3>   RowVector3d;
  typedef Matrix<double, 3, 1>   ColVector3d;
8
  typedef Matrix<int,    3, 3>   Matrix3i;
10 typedef Matrix<double, 3, 3>   Matrix3d;
```

Listing 12.2 **Eigen** fixed-size vector and matrix representation

However, because data types in R are essentially always dynamic and can be resized at any moment, no accessors or conversion functions exist between the fixed-size representation in **Eigen** and the representation in R. All of the interfaces discussed in this chapter use the dynamically sized vectors and matrices introduced next.

12.2.2 Dynamic-Size Vectors and Matrices

Work with data must accommodate changing data sizes, particularly when used interactively or with varying inputs. The core data type for R work with **Eigen** is therefore the type defined as follows:

```
  typedef Matrix<double, Dynamic, 1>         VectorXd;
2 typedef Matrix<double, Dynamic, Dynamic> MatrixXd;

4 typedef Matrix<int, Dynamic, 1>         VectorXi;
  typedef Matrix<int, Dynamic, Dynamic> MatrixXi;
```

Listing 12.3 Eigen dynamic-size vector and matrix representation

with additional variants for rows and column vectors, as well as different underlying scalar representations for `complex` and `float`. The core R access functions involve the types `VectorXd` and `MatrixXd`.

We can revisit the example from the previous section where we now use dynamic vectors and matrices. Note how the initialization is now at run-time using the overloaded << operator.

```
1 Eigen::VectorXd u(3); u << 1,2,3;
  Eigen::VectorXd v(3); v << 4,5,6;
3
  Eigen::MatrixXd m3 = u * v.transpose();
5 double m4 = u.transpose() * v;

7 Rcpp::Rcout <<   "Outer:\n" << m3 << std::endl;
  Rcpp::Rcout <<   "Inner:\n" << m4 << std::endl;
```

Listing 12.4 A simple **Eigen** example using dynamic-size vectors and matrices

The result is of course the same. What about performance differences? More recent versions of **Rcpp** contain a simple helper class `Timer`. It has to be included explicitly as shown below. We can then continue the example and create simple timed loops:

```
  // include header file for timer
2 #include <Rcpp/Benchmark/Timer.h>

4 // start the timer
  const int n = 1000000;
6 Rcpp::Timer timer;
  for(int i=0; i<n; i++) {
8     m1 = x * y.transpose();
      m2 = x.transpose() * y;
10 }
   timer.step("fixed") ;
12
   for(int i=0; i<n; i++) {
14     m3 = u * v.transpose();
       m4 = u.transpose() * v;
16 }
```

```
    timer.step("dynamic");
18
    for(int i=0; i<n; i++) { } // empty loop
20  timer.step( "empty loop" ) ;

22  Rcpp::NumericVector res(timer);
    for (int i=0; i<res.size(); i++) {
24      res[i] = res[i] / n;
    }
26  Rcpp::Rcout << res << std::endl;
```

Listing 12.5 Comparing performance of simple operations between dynamic and fixed size vectors

The result of this comparison of making one million matrix multiplication of an inner and outer product of two short vectors is rather astounding. The Timer class keeps the data in nanoseconds (provided the operating system supports it). By dividing the results by the number of iterations n, we obtain the cost per iteration:

```
       fixed    dynamic  empty loop
2   0.001129  135.464204   0.000256
```

Listing 12.6 Timing results simple operations betweem dynamic and fixed size vectors

The templated code for fixed-size vectors and matrices is barely slower than the empty loop. Without having inspected the generated machine-language code, we assume that the assignment of the nine elements of the outer-product matrix plus the tenth result from the inner-product scalar are replaced by constant assignments—whereas the loop using dynamic data types takes 135 ns per iteration which is in relatively terms much more than the operation implemented with fixed-size data types.

This example, while unrealistic in its simplicity, shows that modern optimizing compilers, combined with template logic, can result in very efficient code as they essentially factor out invariants.

12.2.3 Arrays for Per-Component Operations

C++ matrix libraries overload the operator * such that (conforming) vectors and matrices can by multiplied. This is very useful for the focus on linear algebra and matrix operations and decompositions. However, programmers also often need per-element operations (as is done in R when doing, say, c(1:3) * c(2:4)).

Eigen supports these operations via the Array template classes. In general, there is a one-to-one mapping between the Matrix and Vector classes, and their Array counterparts as shown in Table 12.1.

Where Vector denotes a single dimension, Array uses one X or digit. For Matrix types, two digits are used to fixed-size objects, and XX denotes variable size arrays. The trailing letter still denotes the storage type.

12.2 Eigen classes

Table 12.1 Mapping between **Eigen** matrix and vector types, and corresponding array types

Vector or Matrix object type	Array object type
VectorXd	ArrayXd
Vector3d	Array3d
MatrixXd	ArrayXXd
Matrix3d	Array33d

Conversion between `Matrix`/`Vector` and `Array` are done, respectively, with the `array()` method for the former, and the `matrix()` method for the latter.

12.2.4 Mapped Vectors and Matrices and Special Matrices

The previous sections illustrated the basic vector and matrix representations in **Eigen**, providing either fixed or dynamically sized storage. To interfacing external libraries, or C and C++ arrays, **Eigen** provides another class: `Map`. This approach fits perfectly with the design of **Rcpp** which operates via proxy classes that access the `SEXP` type of the underlying R object. Using such a "mapped object" requires no additional copy upon construction, allowing for efficient passage of objects from R to code using **Eigen** in the same way that the **Rcpp** classes are lightweight.

In general, one uses the desired representation as a template argument for the `Map` classes, leading to, for example, `Eigen::Map< Eigen::MatrixXd > ` in the case of a dynamically sized matrix of type `double`. By deploying the `using` directive to import either the full namespace or selected identifiers, this can be reduced to `Map<MatrixXd>`. It is good practice to declare such mapped object as `const` to prevent accidental changes to the memory content of a mapped variable.

Moreover, **Eigen** also supports operations on sparse matrices. The core class is `SparseMatrix` which offers high performance executing yet low memory usage. It is based on a variant of the Compressed Column (or Row) Storage scheme used by other software libraries operating on sparse matrices. The internal representation uses four compact arrays:

Values   stores the coefficient values of the nonzero elements.
InnerIndices   stores the row (or column) indices of the nonzero elements.
OuterStarts   stores for each column (or row) the index of the first nonzero in the previous two arrays.
InnerNNZs   stores the number of nonzeros of each column (or row).

Here "inner" refers to column vector for a column-major matrix (or a row vector for a row-major matrix) and "outer" refers to the other direction.

Eigen also supports matrices with particular known structure such as symmetric matrices (provided as either a lower- or upper-triangular matrix) or banded matrices. In general, these are provided as "views" which means that while the full size is used, only the relevant portion is accessed during operations.

12.3 Case Study: Kalman Filter Using RcppEigen

Section 10.3 above discussed a simple Kalman filter example and showed its implementation in a simple **C++** class using **Armadillo**. For comparison, we can also implement it using **Eigen**.

```
#include <RcppEigen.h>

using namespace Rcpp;
using namespace Eigen;

class Kalman {
private:
    MatrixXd A, H, Q, R, xest, pest;

public:
    // constructor, sets up data structures
    Kalman() {
        const double dt = 1.0;
        A.setIdentity(6,6);
        A(0,2) = A(1,3) = A(2,4) = A(3,5) = dt;

        H.setZero(2,6);
        H(0,0) = H(1,1) = 1.0;

        Q.setIdentity(6,6);
        R = 1000 * R.Identity(2,2);

        xest.setZero(6,1);
        pest.setZero(6,6);
    }

    // sole member function: estimate model
    MatrixXd estimate(const MatrixXd & Z) {
        unsigned int n = Z.rows(), k = Z.cols();
        MatrixXd Y = MatrixXd::Zero(n,k);
        MatrixXd xprd, pprd, S, B, kalmangain;
        VectorXd z, y;

        for (unsigned int i = 0; i<n; i++) {
            z = Z.row(i).transpose();

            // predicted state and covariance
            xprd = A * xest;
            pprd = A * pest * A.transpose() + Q;

            // estimation
            S = H * pprd.transpose() * H.transpose() + R;
            B = H * pprd.transpose();

            kalmangain = S.ldlt().solve(B).transpose();
```

```
48          // estimated state and covariance
            xest = xprd + kalmangain * (z - H * xprd);
50          pest = pprd - kalmangain * H * pprd;

52          // compute the estimated measurements
            y = H * xest;
54          Y.row(i) = y.transpose();
         }
56       return Y;
      }
58 };
```

Listing 12.7 Basic Kalman filter class in C++ using **Eigen**

Listing 12.7 provides a straightforward adaptation of the previous implementation. The switch from **Armadillo** to **Eigen** consists mostly of

- An obvious change in the header file that is included.
- Switching declarations from mat to MatrixXd, and from vec to VectorXd.
- Changing member functions zero(), identity() and t() to setZero(), setIdentity() and transpose(), respectively.
- Selecting a robust Cholesky decomposition with pivoting matrix decomposition method via the ldlt() member function to use the solve() method.

The code is slightly more verbose than the variant in Listing 10.10. And while **Eigen** has a reputation for providing fast-running code, it turns out that for these (arguably rather naïve) implementations, **Armadillo** holds a considerable speed gain of more than 60 % over **Eigen** using the code shown in Listings 10.10 and 12.7.[2]

12.4 Linear Algebra and Matrix Decompositions

12.4.1 Basic Solvers

Eigen has very substantial support for linear algebra operations and a number of matrix decompositions. Bates and Eddelbuettel (2013) provide a thorough discussion, so rather than enumerating these again, we will highlight a few key elements. The case study in Sect. 12.5 also deploys a number of these in order to evaluate their relative performance in a reimplementation of a linear model estimator. Useful documentation for these methods is provided by the corresponding tutorial section of the Eigen documentation (Guennebaud et al. 2012)[3] from which we have taken the following examples.

---

[2] A comment from another R / **Eigen** developer confirming this ratio is gratefully acknowledged.

[3] The **Eigen** tutorial can be accessed via http://eigen.tuxfamily.org/dox, and more detailed documentation about matrix decompositions is at http://eigen.tuxfamily.org/dox/TopicLinearAlgebraDecompositions.html.

The solvers example can be adapted quite easily to code to be called from R.

```
R> src <- '
   const Map<MatrixXd>   A(as<Map<MatrixXd> >(As));
   const Map<VectorXd>   b(as<Map<VectorXd> >(bs));
   VectorXd x = A.colPivHouseholderQr().solve(b);
   return wrap(x);'
R> solveEx <- cxxfunction(signature(As = "mat", bs = "vec"),
+                         body=src, plugin="RcppEigen")
R> A <- matrix(c(1,2,3,  4,5,6,  7,8,10), 3,3, byrow=TRUE)
R> b <- c(3, 3, 4)
R> solveEx(A, b)
[1] -2  1  1
R>
```

Listing 12.8 Using a basic **Eigen** solver from R

In this example, we pass a matrix and vector from R, and the R data types are used to instantiate the corresponding **Eigen** objects. As discussed in the previous section, a Map type permits us to reuse the R memory without an additional copy of data, and we use the dynamically sized type with double precision. In the example, the matrix A is decomposed using a column-pivoting Householder QR decomposition after which the linear equation

$$Ax = b$$

is solved for a given b.

12.4.2 Eigenvalues and Eigenvectors

Eigenvalues and eigenvector calculation are also available. The following example uses a self-adjoint solver suitable for symmetric matrices in which only one triangle of the corresponding matrix is used, while the other is inferred. Alternate solvers classes EigenSolver and ComplexEigenSolver are also available.

```
R> src <- '
+     using namespace Eigen;
+     const Map<MatrixXd>  A(as<Map<MatrixXd> >(As));
+     SelfAdjointEigenSolver<MatrixXd> es(A);
+     if (es.info() != Success) stop("Problem with Matrix");
+     return List::create(Named("values")  = es.eigenvalues(),
+                         Named("vectors") = es.eigenvectors());'
R> eigEx <- cxxfunction(signature(As = "mat"), body=src,
+                       plugin="RcppEigen")
R> A <- matrix(c(1,2,  2,3), 2,2, byrow=TRUE)
R> eigEx(A)
$values
[1] -0.236068  4.236068

$vectors
```

```
              [,1]       [,2]
 [1,]   -0.850651  -0.525731
 [2,]    0.525731  -0.850651

R>
R> eigEx(matrix(c(1,NA,NA,1),2,2))
Error: Problem with Matrix
R>
```

Listing 12.9 Computing eigenvalues using **Eigen**

Line four shows how a member function of the solver can be queried for success or failure; we then use the stop() wrapper around the **Rcpp** exception handlers to return to R with an appropriate error message. Lines 21 and 22 illustrate this with a degenerate matrix. As expected, control is returned to R with the error message specified in line four.

12.4.3 Least-Squares Solvers

Listing 12.7 in Sect. 12.3 already showed the use of the ldlt() member function for solving linear systems. The following example uses a basic SVD approach. The next section will revisit this problem in more detail.

```
R> src <- '
+     using namespace Eigen;
+     const Map<MatrixXd>  X(as<Map<MatrixXd> >(Xs));
+     const Map<VectorXd>  y(as<Map<VectorXd> >(ys));
+     VectorXd x = X.jacobiSvd(ComputeThinU|ComputeThinV).solve(y);
+     return wrap(x);'
R> lsEx <- cxxfunction(signature(Xs = "matrix", ys = "vector"),
+                     body=src, plugin="RcppEigen")
R> data(cars)
R> X <- cbind(1, log(cars[,"speed"]))
R> y <- log(cars[,"dist"])
R> lsEx(X, y)
[1] -0.729669  1.602391
R>
```

Listing 12.10 Computing least-squares using **Eigen**

We use the standard R data set cars for the well-known regression example of fitting the logarithm of distance to a constant and the logarithm of speed.

12.4.4 Rank-Revealing Decompositions

The **Eigen** library also supports a number of rank-revealing decompositions which can compute the rank of the matrix they are operating on. Such methods tend to be best-behaved in the case of matrices of less than full rank, as, for example, singular matrices in the case of squared dimensions. The reference in footnote 3 on page 183 provides the full details about all available methods.

```
R> src <- '
+    using namespace Eigen;
+    const Map<MatrixXd>  A(as<Map<MatrixXd> >(As));
+    FullPivLU<MatrixXd> lu_decomp(A);
+    return List::create(Named("rank")     = lu_decomp.rank(),
+                        Named("nullSpace") = lu_decomp.kernel(),
+                        Named("colSpace")  = lu_decomp.image(A));
+ '
R> rrEx <- cxxfunction(signature(As = "mat"), body=src, plugin="RcppEigen")
R> A <- matrix(c(1,2,5, 2,1,4, 3,0,3),3,3,byrow=TRUE)
R> rrEx(A)
$rank
[1] 2

$nullSpace
     [,1]
[1,]  0.5
[2,]  1.0
[3,] -0.5

$colSpace
     [,1] [,2]
[1,]    5    1
[2,]    4    2
[3,]    3    3

R>
```

Listing 12.11 Rank-revelaing decompositions using **Eigen**

The example discussed in this section illustrates how fine-grained the **Eigen** API is: a variety of basic decompositions (SVD, LU, QR, ...) can be deployed, and pivoting schemes are available for several of them. The vignette in package **RcppEigen** also provides more detail. The next section provides a more in-depth discussing about how to use these in order to estimate linear models.

12.5 Case Study: C++ Factory for Linear Models in RcppEigen

The **RcppEigen** package continues a theme started by **RcppArmadillo** (François et al. 2012) and **RcppGSL** (François and Eddelbuettel 2010). It consists of taking the venerable linear model estimation as the basis for comparison between different linear algebra implementations. Doug Bates took this a step further with **RcppEigen** by providing a complete "factory" for linear models.

12.5 Case Study: C++ Factory for Linear Models in **RcppEigen**

A "factory," in software engineering parlance, is a set of code, frequently implemented as functions, that produces objects given a set of parameters. Often these objects stem from classes which are related by class inheritances. This is commonly implemented with a base, or top-level, class from which the various models derive. One or more parameters are then used to select and initiate the type of object desired.

In our context, this provides an excellent illustration for both a set of more advanced **C++** code and an opportunity to detail more of the components of **Eigen** and **RcppEigen**. The lm class in Listing 12.12 is the base class from which the factory methods derive.

```
  namespace lmsol {
2   using Eigen::ArrayXd;
    using Eigen::Map;
4   using Eigen::MatrixXd;
    using Eigen::VectorXd;
6
    class lm {
8   protected:
      Map<MatrixXd>        m_X;       // model matrix
10    Map<VectorXd>        m_y;       // response vector
      MatrixXd::Index      m_n;       // number of rows of X
12    MatrixXd::Index      m_p;       // number of columns of X
      MatrixXd::VectorXd   m_coef;    // coefficient vector
14    int                  m_r;       // comp. rank or NA_INTEGER
      MatrixXd::VectorXd   m_fitted;  // vector of fitted values
16    MatrixXd::VectorXd   m_se;      // standard errors
      MatrixXd::RealScalar m_prescribedThreshold;
18                                    // user specified tolerance
      bool                 m_usePrescribedThreshold;
20
    public:
22    lm(const Map<MatrixXd>&, const Map<VectorXd>&);

24    ArrayXd              Dplus(const ArrayXd& D);
      MatrixXd             I_p() const {
26        return MatrixXd::Identity(m_p, m_p);
      }
28    MatrixXd             XtX() const;

30    // setThreshold + threshold based on ColPivHouseholderQR
      RealScalar           threshold() const;
32    const VectorXd&           se() const {return m_se;}
      const VectorXd&         coef() const {return m_coef;}
34    const VectorXd&       fitted() const {return m_fitted;}
      int                     rank() const {return m_r;}
36    lm&                 setThreshold(const RealScalar&);
    };
38
    // ..
40 }
```

Listing 12.12 Core of definition of lm class in **Eigen**

The implementation of the non-inlined member functions is provided in the package **RcppEigen** as the source file fastLm.cpp. We will omit these functions here due to space constraints.

With the declarations of the basic linear model class lm, we can define specializations providing the various decompositions. As these classes all inherit from lm, they share all its member functions and variables shown in Listing 12.12 yet each add their own specific decomposition function—simply by instantiating the corresponding class from **Eigen**.

In the implementation shown below in Listing 12.13, the classes deriving from the lm class shown above all instantiate an **Eigen** object of the same name. This is made possible by the different namespaces. The Eigen namespace is used by the **Eigen** package (and we have omitted a number of statements such as using Eigen::Llt which permit use of the Llt class without the namespace prefix), and the lmsol namespace is used for the "linear model solutions" implemented in this example from the **RcppEigen** package.

So to take the first example, the ColPivQR class in the lmsol namespace inherits from lm in the same namespace and provides access to the Eigen::ColPivQR class from **RcppEigen**. We sometimes prefer to write this explicitly—and the two forms lmsol::ColPivQR and Eigen::ColPivQR make the provenance more explicit.

```
    class ColPivQR : public lm {
    public:
        ColPivQR(const Map<MatrixXd>&, const Map<VectorXd>&);
    };

    class Llt : public lm {
    public:
        Llt(const Map<MatrixXd>&, const Map<VectorXd>&);
    };

    class Ldlt : public lm {
    public:
        Ldlt(const Map<MatrixXd>&, const Map<VectorXd>&);
    };

    class QR : public lm {
    public:
        QR(const Map<MatrixXd>&, const Map<VectorXd>&);
    };

    class GESDD : public lm {
    public:
        GESDD(const Map<MatrixXd>&, const Map<VectorXd>&);
    };

    class SVD : public lm {
    public:
        SVD(const Map<MatrixXd>&, const Map<VectorXd>&);
    };

```

12.5 Case Study: C++ Factory for Linear Models in **RcppEigen**

```
        class SymmEigen : public lm {
32      public:
            SymmEigen(const Map<MatrixXd>&, const Map<VectorXd>&);
34      };
```

Listing 12.13 Derived classes of lm providing specializations

With these declarations (and the actual implementations which are included in the **RcppEigen** package as file `fastLm.cpp`), we can show the implementation of the C++ part of the `fastLm()` function. But before we go there, we will illustrate two of the different constructors.

```
1  QR::QR(const Map<MatrixXd> &X,
            const Map<VectorXd> &y) : lm(X, y) {
3      HouseholderQR<MatrixXd> QR(X);
       m_coef    = QR.solve(y);
5      m_fitted  = X * m_coef;
       m_se      = QR.matrixQR().topRows(m_p).
7                    triangularView<Upper>().
                     solve(I_p()).rowwise().norm();
9  }

11 Llt::Llt(const Map<MatrixXd> &X,
            const Map<VectorXd> &y) : lm(X, y) {
13     LLT<MatrixXd>  Ch(XtX().selfadjointView<Lower>());
       m_coef    = Ch.solve(X.adjoint() * y);
15     m_fitted  = X * m_coef;
       m_se      = Ch.matrixL().solve(I_p()).colwise().norm();
17 }
```

Listing 12.14 Implementation of two subclass constructors for lm model fit

These two examples show that particular aspects of the respective **Eigen** classes are used. For the QR decomposition variant of the linear model, the coefficients are provided via `solve()` to obtain the parameter vector. Fitted values are then just a multiplication with the original design matrix, and standard errors can be computed exploiting properties of the QR decomposition. This is similar for the Llt approach; the full source of `fastLm.cpp` in the package **RcppEigen** provides the full detail.

Before we can address the implementation of the actual linear model fit, we first define an inlined helper function. It creates the corresponding object given the matrix X, vector y, and a variable named `type` to select the given "type" of decomposition used for the model fit:

```
1  static inline lm do_lm(const Map<MatrixXd> &X,
                          const Map<VectorXd> &y,
3                         int type) {
       switch(type) {
5      case ColPivQR_t:
           return ColPivQR(X, y);
7      case QR_t:
           return QR(X, y);
9      case LLT_t:
           return Llt(X, y);
```

```
         case LDLT_t:
             return Ldlt(X, y);
         case SVD_t:
             return SVD(X, y);
         case SymmEigen_t:
             return SymmEigen(X, y);
         case GESDD_t:
             return GESDD(X, y);
         }
         throw invalid_argument("invalid type");
         return ColPivQR(X, y);   // -Wall
     }
```

Listing 12.15 Selection of subclasses for lm model fit

Note that as the do_lm function is in the lmsol namespace, it does instantiate the subclasses of lm declared in Listing 12.13 rather than the **Eigen** classes they provide access to.

Finally, the actual linear model function called from R:

```
     extern "C" SEXP fastLm(SEXP Xs, SEXP ys, SEXP type) {
       try {
         const Map<MatrixXd>  X(as<Map<MatrixXd> >(Xs));
         const Map<VectorXd>  y(as<Map<VectorXd> >(ys));
         Index                n = X.rows();
         if ((Index)y.size() != n)
             throw invalid_argument("size mismatch");

         // Select and apply the least squares method
         lm                   ans(do_lm(X, y, ::Rf_asInteger(type)));

         // Copy coefficients and install names, if any
         NumericVector    coef(wrap(ans.coef()));
         List             dimnames(NumericMatrix(Xs).attr("dimnames"));
         if (dimnames.size() > 1) {
           RObject    colnames = dimnames[1];
           if (!(colnames).isNULL())
             coef.attr("names") = clone(CharacterVector(colnames));
         }

         VectorXd         resid = y - ans.fitted();
         int              rank = ans.rank();
         int    df = (rank == ::NA_INTEGER) ? n - X.cols() : n - rank;
         double s = resid.norm() / std::sqrt(double(df));
                            // Create the standard errors
         VectorXd         se = s * ans.se();

         return List::create(_["coefficients"] = coef,
                             _["se"]           = se,
                             _["rank"]         = rank,
                             _["df.residual"]  = df,
                             _["residuals"]    = resid,
                             _["s"]            = s,
                             _["fitted.values"] = ans.fitted());
```

12.5 Case Study: C++ Factory for Linear Models in **RcppEigen**

```
      } catch( std::exception &ex ) {
36         forward_exception_to_r( ex );
      } catch(...) {
38         ::Rf_error( "c++ exception (unknown reason)" );
      }
40    return R_NilValue; // -Wall
}
```

Listing 12.16 Actual `fastLm` function in `RcppEigen` package

Here the `ans` object is instantiated with the return from the `do_lm` function from the preceding listing. This `ans` object then provides the appropriate solutions given the type of decomposition chosen. Together, this implements a very elegant setup providing a large number of different approaches (which can then be compared) with a minimal amount of code repetition. This illustrates nicely how C++ design choices enable us to provide code very effectively from R while also computing efficiently thanks to the advanced features in **Eigen**.

The package vignette (Bates et al. 2012) has the complete details, but we can restate the result of the comparison computed with the help of the code listings shown above (as well as the rest not shown here but available in the **RcppEigen** package).

All solutions referenced in Table 12.2 refer to the corresponding **Eigen** classes as implemented in the `fastLm` function in the **RcppEigen** package. Exceptions are "arma" and "GSL" for the corresponding `fastLm` functions from the **RcppArmadillo** and **RcppGSL** packages, and "lm.fit" for the base R function.

Table 12.2 `lmBenchmark` results for the **RcppEigen** example

Method	Relative	Elapsed	User	Sys
LDLt	1.000	4.423	4.388	0.020
LLt	1.003	4.438	4.389	0.032
SymmEig	2.629	11.629	10.253	1.320
QR	5.117	22.631	21.205	1.340
arma	5.215	23.068	77.020	15.045
PivQR	5.502	24.335	22.477	1.772
lm.fit	6.086	26.919	45.143	50.951
GESDD	9.582	42.379	126.832	39.782
SVD	33.932	150.082	145.781	3.753
GSL	115.522	510.955	601.682	701.116

The timings are from a desktop computer running the default size, $100,000 \times 40$, full-rank model matrix running 100 repetitions for each method. Times (Elapsed, User and Sys) are in seconds. The BLAS in use is the version of the OpenBLAS library included with Ubuntu 12.10. The processor used for these timings is a 4-core processor but almost all the methods are single-threaded and not affected by the number of cores. Only the `arma`, `lm.fit`, `GESDD`, and `GSL` methods benefit from the multi-threaded **BLAS** implementation provided by **OpenBLAS**.

What we can take away from these results is that methods based on forming and decomposing $X^\top X$ (LDLt, LLt and SymmEig) are considerably faster.

The pivoted QR method is marginally times faster than `lm.fit` (from R) on this test and provides nearly the same information as `lm.fit` (which has improved its performance relative to older versions of R). Methods based on the singular value decomposition (SVD and GSL) are much slower, which is presumably caused at least in part by X having many more rows than columns. Also, the GSL method from the GNU Scientific Library uses an older algorithm for the SVD and is clearly not competitive in this comparison.

We also note that `GESDD` implements an interesting hybrid approach by using **Eigen** classes, but calling out to the LAPACK routine `dgesdd` for the actual SVD calculation. This leads to better performance compared to using the SVD implementation of **Eigen** which, while not as bad as the GSL, is still not a particularly fast SVD method.

This example, developed by Doug Bates and implemented as an example in the **RcppEigen** package, provided a nice illustration about the potential for a **Rcpp**-based solution to accelerate computations done in R. **Rcpp** permits us to connect to modern linear algebra libraries such as **Armadillo** (Sanderson 2010) and **Eigen** (Guennebaud et al. 2012) with ease. As can be seen from Table 12.2, a sizeable improvement can be achieved even against the fastest and purest solution offered by R, even if that solution is already fairly efficient and itself implemented mostly in compiled code as is the case with `lm.fit()`, the core function underlying linear model estimation in R.

As of early 2013, over 100 on CRAN and 10 packages on BioConductor use **Rcpp** and offer a wide variety of different choices of how to enhance R seamlessly with C++. The `rcpp-devel` mailing list is vibrant, development of **Rcpp** continues at a rapid pace, and we look forward to more exciting activity in seamlessly bridging R and C++.

Part V
Appendix

Appendix A
C++ for R Programmers

Abstract The short appendix offers a very basic introduction to the C++ language to someone already (at least somewhat) familiar with R programming. Introducing all of C++ in just a few pages is not really possible. Countless books have been written about the C++ language since its inception in the early 1990s (and we will list a few at the end in a section on further readings).

A.1 Compiled Not Interpreted

One of the key differences between R and C++ is that R is interpreted. It was designed for interactive exploration, visualization, and modeling. The flexibility that such a goal aspires to is most naturally reflected in a language with features common to those held by R. This includes "computing on the language" with objects which modify other objects, or functions and more.

C++, on the other hand, came after C and has always been *compiled*. This means that we take a file containing *source code* and convert it into *object code* with a compiler. A linker then creates an executable out of the object code as well as potentially further libraries on the system. We should also note that R itself now ships with a (byte-code) compiler, but that is slightly different as it generates an intermediate level of parsed expressions, rather than machine code in object files as a standard compiler for a language such as C or C++ would.

Let us consider a concrete example. If the code below

```
#include <cstdio>

int main(void) {
    printf("Hello, World!\n");
}
```

Listing A.1 Simple C++ example: Hello, World!

is saved in a file `ex1.cpp`, then the commands

```
sh> g++ -c ex1.cpp
sh> g++ -o ex1 ex1.o
```

Listing A.2 Compiling and linking simple C++ example: Hello, World!

first compile the source file into the object file `ex1.o` as requested by the `-c` command-line option. Next, the resulting object file is linked into the executable `ex1` as specified by the `-o` command-line argument.

This can also be achieved in a single operation via

```
sh> g++ ex1.cpp -o ex1
```

Listing A.3 Compiling and linking simple C++ example in one step: Hello, World!

The resulting program `ex1` can now be executed. It displays the text that is ever so common for first examples by calling the C-level function `printf` which may be somewhat familiar to R programmers via the related R function `sprintf` which uses similar formatting rules to print into a character variable. Notice how we also specified a so-called *include file* in the first line; it contains a number of function declarations related to input and output such as `printf`.

These operations would be the same on any operating system on which g++ has been installed, in particular Windows, OS X, or Linux. The file extensions used for object files or executable may differ, the compile commands remain the same. As an aside, such portability of tools across operating system is a very useful attribute which the R software system shares.

Other compilers can also be used for as long as they are supported by R itself. As noted in Chap. 2 above, this excludes the Microsoft family of compilers, but may include the (commercial) Intel Compiler on several platforms, the Sun compiler if installed on Solaris or Linux, and older Unix compilers such as the IBM AIX compiler and the HP UX compiler. However, as such operating systems are less commonly used, we will concentrate on g++.

An important second aspect of compilation concerns how to build upon other projects via their libraries (providing code) and header files (providing declarations). For example, the R environment provides a stand-alone library with several of the mathematical, probability, and random-number functions used in R (R Development Core Team 2012d, Section 16.6).

Consider the following example which uses the R stand-alone mathematics library to compute the 95 % percentile of the $N(0,1)$ distribution.

```
#include <cstdio>
#include <Rmath.h>

int main(void) {
  printf("N(0,1) 95th percentile %9.8f\n",
      qnorm(0.95, 0.0, 1.0, 1, 0));
}
```

Listing A.4 Simple C++ example using **Rmath**

We can build this program via

```
sh> g++ -c ex2.cpp -DMATHLIB_STANDALONE -I/usr/include
sh> g++ -o ex2 ex2.o -L/usr/lib -lRmath
```

Listing A.5 Compiling and linking simple C++ example using **Rmath**

which shows two new aspects in the compile step. First, we inform the compiler where to find the header file Rmath.h (which contains the declarations) by using the -I/usr/include switch. We define a variable MATHLIB_STANDALONE to enable the stand-alone use of the library outside of its normal deployment with the R engine. Next, for the linking step, the -L/usr/lib switch points to the library location whereas -lRmath enables linking with the R mathematics library from the file libRmath.so (or libRmath.a in case of static linking). In this particular case both the header file location and the library location actually correspond to system defaults. This means we could have omitted them both; however, it is instructive to show them in case the location does need to be specified as would be the case, say, with a local installation in the home directory of the user.

Understanding compiling and linking options and common error messages is of some importance when working with compiled code. For most of our cases, R helps by providing complete wrappers via sub-commands such as R CMD COMPILE or R CMD LINK. However, it is helpful to understand basic compiling and linking in order to examine or debug possible build issues.

A.2 Statically Typed

A second key difference between R and C++ concerns the difference between dynamic and static typing. In R, an expression determines the type it is assigned to. In other words, in

```
R> x <- rnorm(10)
R> x <- "some text"
```

Listing A.6 Simple R example of dynamic types

the variable x is first assigned a numeric (or floating-point) vector of size ten as returned from the rnorm function. This value in x is then replaced by characters as a fixed text is assigned. This is completely valid R code where the result of the expression determines the type of variable it is assigned to: dynamic typing.

Statically typed languages such as C or C++ are different. Variables have to be declared first which assigns the name of a variable to a particular type. That type is then fixed for as long as this variable is *in scope* which may be as long as the program runs, or just a fraction of a second until the current scope (typically defined by a pair of curly braces) is exited. A certain number of assignments from one type to another are possible. For example, assigning a floating number such as 3.1415 to an integer truncates (rather than rounds) its value to 3. Assigning it back to a floating-point variable would then make it 3.0. In other words, the assignment may (or may not) be losing precision, and it depends on the type of variable assigned to and from.

Standard variable types in C++ are

- Integer of different sizes and hence supported ranges of values; int and long are most common; they can also be qualified as unsigned which excludes negative values and thereby doubles the range of positive values.
- Floating point numbers of lower (float) and higher (double) precision.
- Logical values such as bool.
- Character values as char but these are individual letters or symbols, not compounds such as *strings* as there is no base type for strings (but see below for the STL strings).

Another key difference is that all these types are scalar. Vectors can be created statically with size fixed at compile time, or dynamically as in C. That is, however, a feature which can be avoided almost entirely by relying on STL types as discussed below.

A.3 A Better C

C++ can also be seen as better C. In fact, Meyers (2005) argues in his first of 55 "items" that C++ should be seen as a federation of four languages, with C being one of these. Hence, we need to review a few core language elements which are actually fairly similar to the R language.

Control Structures

C++ contains several control structures which are similar to those in R:

- for loops are very common; they contain three components *initialization, comparison for termination*, and *incrementing*. So in

```
  for (int i=0; i<10; i++) {
2   // some code here
  }
```

Listing A.7 Simple R example of dynamic types

the loop body will be entered ten times with the variable i ranging from zero to nine. Once the expression $i < 10$ no longer evaluates to true, the code resumes after the end of the for block.
- while loops are also similar with a top-level boolean expression and a loop body that is entered for as long as the condition is true; a related but much less used variant starts with the do keyword and the loop body and the test at the end; lastly, keywords break and next exist to exit the loop body and skip to the next iterations, respectively.

A.3 A Better C

- `if` statements are very similar to what one uses in R with optional `else` blocks and nesting; very little is different here.
- `switch` statements are an alternative to "ladders" of `if`/`else` as a single statement evaluates and the matching condition, represented by a `case` label, is executed, or else a default value is chosen.

Functions

Functions also share some similarities with their R equivalents. Functions can be defined to take a number of arguments. Argument matching is always by position; passing arguments by name as in R is not permitted. All listed function arguments have to be supplied. An example of this was the `qnorm` function above which we had to call with all five arguments. Its R version also has up to five arguments, but if called as `qnorm(0.95)`, default values for mean, standard deviation, lower tail, and logarithmic use apply. In C++, we explicitly list all five arguments (though default arguments can be supplied as well in a function definition).

Because the language is statically typed, functions are differentiated by both their names and argument types. That means that these two function declarations

```
int    min(int a, int b);
double min(double a, double b);
```

Listing A.8 Simple C++ function example

are in fact distinct. The compiler will call the corresponding ones for `min(4, 5)` and `min(4.0, 5.0)`. Templates, discussed below, offer an approach to write more generic functions that apply to several variable types.

Pointers and Memory Management

Pointers and memory management is an important advanced topic, particularly for C programming. In C++, use of pointers can in many cases be avoided, diverting one frequent case of criticism.

There are two very common use cases for pointers. The first one concerns dynamic memory allocation. In C, the only approach to reserve a vector or array (of, say, type `double`) of a size given only at run-time is to declare a pointer to double. At run-time, when the required size is known, this pointer is then assigned a dynamic memory allocation of the appropriate size determined as the number of required elements times the size of a double. After use, the memory has to be freed or else it *leaks* which means it is allocated but not used; a waste of system resources. This process sounds manual and error-prone, and it is. But with C++ and facilities such as the Standard Template Library (STL) discussed below, we do not need to resort to this approach to have dynamically sized vectors or arrays.

The second aspect concerns how arguments are passed to functions. There are two approaches in C and by extension C++. The first is *call-by-value* in which a copy is passed to the subroutine. It can be modified at will and changes will not affect the calling function and its value. That is safe yet occasionally inefficient (as larger compound data types will be copied in full) or even inapplicable. The second use case is *call-by-reference*. Here, a pointer is passed and one can access the original memory location. This is frequently more efficient, and also the only way to alter an object. C++ improves upon this setup by offering a call-by-reference without pointers.

```
#include <cstdio>

void abs(double & x) {
   if (x < 0)
      x = -x;
}

int main(void) {
   double x = -3.4;
   printf("%f\n", x);
   abs(x);
   printf("%f\n", x);
}
```

Listing A.9 Simple C++ function call example

Here a function which changes its argument to its absolute value is defined and tested. The output is first negative, and then positive. No pointers are used, yet the value is changed as the variable is passed by reference, indicated by the & in the function signature. (The example is contrived, typically we write a function for an absolute value as returning the modified value, rather than changing the argument.)

A.4 Object-Oriented (But Not Like S3 or S4)

The second of the four languages "federated inside Cpp" according to Meyers (2005) is object-oriented C++. There is an enormous amount of complexity around both the *how* and *why* of object-oriented programming, both in general and specifically in C++ which is arguably a fairly complex language. That said, some high-level concepts are easy enough to express in a few paragraphs, and we will concentrate on such a higher-level approach here.

The basic composite type in C++ is a `struct`, which is inherited from C. It offers the most basic form of composition as it permits to group several variables inside a newly defined type.

```
struct Date {
   unsigned int year
   unsigned int month;
   unsigned int date;
```

```
5 };

7 struct Person {
      char firstname[20];
9     char lastname[20];
      struct Date birthday;
11    unsigned long id;
   };
```

Listing A.10 Simple C++ data structure using `struct`

Here, we define a `Date` type containing year, month, and date as separate unsigned integers. That structure is then reused in the `Person` structure. So far, so good. What is not to like? First, all data elements are by default *public* meaning every piece of code having access to the structure has the ability to change values. Second, the structure really only holds data but no code.

The `class` data type overcomes both by associating *methods* (which are class-specific functions) with the class. Moreover, data can now be *public* (visible to all), *private* (visible only to methods of the class), or *protected* (a refinement having to do with inheritance we can ignore here).

A possible sketch of a class declaration for a `Date` could be

```
  class Date {
2 private:
      unsigned int year
4     unsigned int month;
      unsigned int date;
6 public:
      void setDate(int y, int m, int d);
8     int getDay();
      int getMonth();
10    int getYear();
  }
```

Listing A.11 Simple C++ data structure using `class`

This class contains a few changes as discussed above. Date fields are now private: data cannot be accessed directly from outside the class. To do so, we now have accessor functions. The first sets the date—and, in doing so, can ascertain that the date supplied is actually a valid date. This is followed by three more functions which access the date components. The implementation of the function bodies would then be supplied in a matching `cpp` file complementing the declaration from a header file.

A.5 Generic Programming and the STL

The STL provides the third distinct language aspect within the "federation of four languages" view described by Meyers (2005). The STL has become a staple of C++ programming for efficient and *generic* programming (Austen 1999). In this context, generic means a consistent interface that is provided irrespective of the chosen data type.

As one illustrative example, consider the so-called sequence container types `vector`, `deque`, and `list`. Each of these supports common functions such as

`push_back()`   to insert at the end
`pop_back()`   to remove from the front
`begin()`   returning an iterator to the first element
`end()`   returning an iterator to just after the last element
`size()`   for the number of elements

and more similar functions. However, `list` offers different performance guarantees and implementation details than `vector` so it can also offer complementary functions such as `push_front()` and `pop_front()` which a vector does not have. On the other hand, `v[i]` can access the element with index *i* in the vector (which offers random access), whereas a list has to be traversed. The `deque` class provides aspects of both `vector` and `lists` and can be seen as a compromise or superset of both features sets.

Other commonly used container types are associative:

`set`   for collections of objects where both key and value are the same; it provides set-theoretic operations such as union and intersect.
`multiset`   which extends `set` by allowing several instances of the same key/value.
`map`   which is pair associative container linking a key to a value; both can be of different types mapping, e.g., a numeric index (e.g., a zip or postal code represented as an integer) to a character vector or string type with the name of the municipality.
`multimap`   which extends `map` by allowing an unlimited number of values for a given key.

as well as hashed versions of these types. In the original SGI implementation of the STL these are named `hash_*` but the name `ordered_*` was chosen in the current TR1 implementation of the upcoming C++ standard.

One commonality of both sequence and associative containers is the traversal via `iterators`. Consider this example for the `vector` class where we use the `const_iterator` variant which indicates that it accesses elements read-only but never modifies:

```
    std::vector<double>::const_iterator si;
    for (si=s.begin(); si != s.end(); si++) {
        std::cout << *si << std::endl;
    }
```

Listing A.12 Simple C++ example using iterators on `vector`

We can use the exact same `for` loop simply by modifying the iterator to be a `list` type

```
    std::list<double>::const_iterator si;
```
Listing A.13 Simple C++ example using const iterators on `list`

or change it to the `deque` type

```
1   std::deque<double>::const_iterator si;
```
Listing A.14 Simple C++ example using const iterators on `deque`

which illustrates the *generic* nature of STL operators.

The STL also contains a number of algorithms. A simple one is *accumulate* which can be used as

```
1   std::cout << "Sum is "
            << std::accumulate(s.begin(), s.end(), 0)
3           << std::endl;
```
Listing A.15 Simple C++ example using `accumulate` algorithm

and this is irrespective of what class the object s is instantiated from for as long as it supports iterator access as well as `begin()` and `end()`. The third argument is the initial value of the summation which we set to zero.

Other popular STL algorithms are

find   which finds the first element equal to the supplied value, if any.
count   which counts the number of elements match the given condition.
transform   which applies a supplied unary or binary function to each element.
for_each   which also sweeps over all elements but does not alter elements.
inner_product   which can be used to compute the inner product of two vectors, or a sum of squares of a single vector.

The key insight here is that these algorithms and iterators can be applied to the different data structures—sequential containers as well as associative containers—with very minimal change. It is in this sense that programming with the STL is "generic."

Many more algorithms are available and described, for example, in Meyers (2001). A final note is that with the STL now being part of the language standard, the use of the term "STL" which refers to what was once an external library is no longer entirely correct. The Standard C++ Library is more appropriate. However, it is still common to refer to these parts of the library as STL reflecting their historical source of having been an extension to the then-smaller standard library.

A.6 Template Programming

Template programming provides the fourth and last language federated within C++ as per Meyers (2005). It is arguably the most complex aspect and the one in which C++ differs most from other related languages such as Java or C#.

Templates programming and its use can range from the very simple to the very complex. Examples for the more complex end of template use are provided by the template meta-programming technique. It is at the core of both Armadillo (Sanderson 2010), Eigen (Guennebaud et al. 2012), and *Rcpp sugar* discussed in Chap. 8.

This section, however, will focus on simpler uses of templates. An example above considered different `min` functions for integers and doubles. A more general solution uses templates:

```
template <typename T>
const T& min(const T& x, const T& y) {
    return y < x ? y : x;
}
```

Listing A.16 Simple C++ template example

This returns a constant reference of the templated type `T` which is also used for the input types. The sole expression uses the standard C comparison operator to return the smaller of the two arguments `x` and `y`.

A simple example of template use was already shown in Sect. 2.5.2. That section illustrates how to combine the **inline** package with short text and code snippets similar to header files. As a concrete example, the following templated class that squares its input was shown:

```
template <typename T>
class square : public std::unary_function<T,T> {
public:
    T operator()( T t) const {
        return t*t;
    }
};
```

Listing A.17 Another C++ template

Templates are used throughout the **Rcpp** sources. Key components of **Rcpp** such as the conversion function `as<>()` are implemented using templates. The `as<>()` conversion function accepts a template type and converts a `SEXP` provided as input to this type (provided the conversion is suitable; else an exception is thrown). However, the inverse operation, provided by `wrap`, is a standard function which does not use templates: the dispatch is done on the basis of the main argument type.

Template programming is a more advanced form of C++ use. It can get rather complicated rather quickly; so we will not dive deeper into templates but refer the reader to the literature.

A.7 Further Reading on C++

A standard reference and introduction to C++ is provided by the creator of the language in Stroustrup (1997). This book is generally not recommended as a first book on C++ for which Lippman et al. (2005) is more frequently listed. Meyers (2005,

1995, 2001) is a highly recommended and readable series of "items" suggesting best practices for C++ and STL use.

As C++ is a popular and widely used programming language, several good resources exist on the Internet as well. The Wikipedia page[1] provides a very good start with numerous further references. Brokken (2012) is a recommended and freely downloadable text introducing C++ which has been maintained and extended since 1994. Also, good introductions to template programming are provided by Abrahams and Gurtovoy (2004) and Vandevoorde and Josuttis (2003).

Finally, among C++ projects, **Boost** (at http://www.boost.org) stands out and deserves special mention. Boost is a collection of several dozen rigorously developed and peer-reviewed libraries. Some of the **Boost** libraries will be included in the next version of the C++ standard.

---

[1] See http://en.wikipedia.org/wiki/C++.

References

Abrahams D, Grosse-Kunstleve RW (2003) Building Hybrid Systems with Boost.Python. Boost Consulting, URL http://www.boostpro.com/writing/bpl.pdf

Abrahams D, Gurtovoy A (2004) C++ Template Metaprogramming: Concepts, Tools and Techniques from Boost and Beyond. Addison-Wesley, Boston

Adler D (2012) rdyncall: Improved Foreign Function Interface (FFI) and Dynamic Bindings to C Libraries. URL http://CRAN.R-Project.org/package=rdyncall, R package version 0.7.5

Albert C, Vogel S (2012) GUTS: Fast Calculation of the Likelihood of a Stochastic Survival Model. URL http://CRAN.R-Project.org/package=GUTS, R package version 0.2.8

Armstrong W (2009a) RAbstraction: C++ abstraction for R objects. URL http://github.com/armstrtw/rabstraction, code repository last updated July 22, 2009.

Armstrong W (2009b) RObjects: C++ wrapper for R objects (a better implementation of RAbstraction). URL http://github.com/armstrtw/RObjects, code repository last updated November 28, 2009.

Auguie B (2012a) cda: Couple dipole approximation. URL http://CRAN.R-Project.org/package=cda, R package version 1.2.1

Auguie B (2012b) planar: Multilayer optics. URL http://CRAN.R-Project.org/package=planar, R package version 1.2.4

Austen MH (1999) Generic Programming and the STL: Using and Extending the C++ Standard Template Library. Addison-Wesley

Bates D, DebRoy S (2001) C++ classes for R objects. In: Hornik K, Leisch F (eds) Proceedings of the 2nd International Workshop on Distributed Statistical Computing (DSC 2001), TU Vienna, Austria

Bates D, Eddelbuettel D (2013) Fast and elegant numerical linear algebra using the RcppEigen package. Journal of Statistical Software 52(5), URL http://www.jstatsoft.org/v52/i05

Bates D, François R, Eddelbuettel D (2012) RcppEigen: Rcpp integration for the Eigen templated linear algebra library. URL http://CRAN.R-Project.org/package=RcppEigen, R package version 0.3.1.2

Brokken FB (2012) C++ annotations. Electronic book, University of Groningen, URL http://www.icce.rug.nl/documents/cplusplus/, version 9.4.0, accessed 2012-11-24.

Chambers JM (1998) Programming with Data: A Guide to the S Language. Springer, Heidelberg, ISBN 978-0387985039

Chambers JM (2008) Software for Data Analysis: Programming with R. Statistics and Computing, Springer, Heidelberg, ISBN 978-0-387-75935-7

Chambers JM, Hastie TJ (1992) Statistical Models in S. Chapman & Hall, London

Eddelbuettel D (2012a) RcppCNPy: Rcpp bindings for NumPy files. URL http://CRAN.R-Project.org/package=RcppCNPy, R package version 0.2.0

Eddelbuettel D (2012b) RcppDE: Global optimization by differential evolution in C++. URL http://CRAN.R-Project.org/package=RcppDE, R package version 0.1.1

Eddelbuettel D, François R (2012a) Rcpp: Seamless R and C++ Integration. URL http://CRAN.R-Project.org/package=Rcpp, R package version 0.10.0

Eddelbuettel D, François R (2012b) RcppBDT: Rcpp binding for the Boost Date_Time library. URL http://CRAN.R-Project.org/package=RcppBDT, R package version 0.2.1

Eddelbuettel D, François R (2012c) RcppClassic: Deprecated 'classic' Rcpp API. URL http://CRAN.R-Project.org/package=RcppClassic, R package version 0.9.2

Eddelbuettel D, François R (2012d) RInside: C++ classes to embed R in C++ applications. URL http://CRAN.R-Project.org/package=RInside, R package version 0.2.7

Eddelbuettel D, Nguyen K (2012) RQuantLib: R interface to the QuantLib library. URL http://CRAN.R-Project.org/package=RQuantLib, R package version 0.3.9

Eddelbuettel D, Sanderson C (2013) RcppArmadillo: Accelerating R with high-performance C++ linear algebra. Computational Statistics and Data Analysis (in press)

Fellows I (2012) wordcloud: Word clouds. URL http://CRAN.R-Project.org/package=wordcloud, R package version 2.2

François R (2012a) highlight: Syntax highlighter. URL http://CRAN.R-Project.org/package=highlight, R package version 0.3–2

François R (2012b) parser: Detailed R source code parser. URL http://CRAN.R-Project.org/package=parser, R package version 0.1

François R, Eddelbuettel D (2010) RcppGSL: Rcpp integration for GNU GSL vectors and matrices. URL http://CRAN.R-Project.org/package=RcppGSL, R package version 0.2.0

François R, Eddelbuettel D, Bates D (2012) RcppArmadillo: Rcpp integration for Armadillo templated linear algebra library. URL http://CRAN.R-Project.org/package=RcppArmadillo, R package version 0.3.4.4

Galassi M, Davies J, Theiler J, Gough B, Jungman G, Alken P, Booth M, Rossi F (2010) GNU Scientific Library Reference Manual. 3rd edn, URL http://www.gnu.org/software/gsl, version 1.14. ISBN 0954612078

Gentleman R (2009) R Programming for Bioinformatics. Computer Science and Data Analysis, Chapman & Hall/CRC, Boca Raton, FL

Gropp W, Lusk E, Doss N, Skjellum A (1996) A high-performance, portable implementation of the MPI message passing interface standard. Parallel Computing 22(6):789–828, URL http://dx.doi.org/10.1016/0167-8191(96)00024-5

Gropp W, Lusk E, Skjellum A (1999) Using MPI: Portable Parallel Programming with the Message Passing Interface, 2nd edn. Scientific and Engineering Computation Series, MIT Press, ISBN 978-0-262-57132-6

Guennebaud G, Jacob B, et al (2012) Eigen v3. URL http://eigen.tuxfamily.org

Hankin RKS (2011) gsl: Wrapper for the Gnu Scientific Library. URL http://CRAN.R-Project.org/package=gsl, R package version 1.9-9

Java JJ, Gaile DP, Manly KE (2007) R/Cpp: Interface classes to simplify using R objects in C++ extensions, URL http://sphhp.buffalo.edu/biostat/research/techreports/UB_Biostatistics_TR0702.pdf, unpublished manuscript, University at Buffalo

Jurka TP, Tsuruoka Y (2012) maxent: Low-memory Multinomial Logistic Regression with Support for Text Classification. URL http://CRAN.R-Project.org/package=maxent, R package version 1.3.2

King M, Diaz FC (2011) RSofia: Port of sofia-ml to R. URL http://CRAN.R-Project.org/package=RSofia, R package version 1.1

Kusnierczyk W (2012) rbenchmark: Benchmarking routine for R. URL http://CRAN.R-Project.org/package=rbenchmark, R package version 1.0

Leisch F (2008) Tutorial on Creating R Packages. In: Brito P (ed) COMPSTAT 2008 – Proceedings in Computational Statistics, Physica Verlag, Heidelberg, Germany, URL http://CRAN.R-Project.org/doc/contrib/Leisch-CreatingPackages.pdf

Liang G (2008) rcppbind: A template library for R/C++ developers. URL http://R-Forge.R-Project.org/projects/rcppbind, R package version 1.0

Lippman SB, Lajoie J, Moo BE (2005) The C++ Primer, 4th edn. Addison-Wesley

Matloff N (2011) The Art of R Programming: A Tour of Statistical Software Design. No Starch Press, San Francisco, CA

Meyers S (1995) More Effective C++: 35 New Ways to Improve Your Programs and Designs. Addison-Wesley Longman Publishing Co., Inc., Boston, MA, USA, ISBN 020163371X

Meyers S (2001) Effective STL: 50 specific ways to improve your use of the standard template library. Addison-Wesley Longman Ltd., Essex, UK, ISBN 0-201-74962-9

Meyers S (2005) Effective C++: 55 Specific Ways to Improve Your Programs and Designs, 3rd edn. Addison-Wesley Professional, ISBN 978-0321334879

R Development Core Team (2012a) R Installation and Administration. R Foundation for Statistical Computing, Vienna, Austria, URL http://CRAN.R-Project.org/doc/manuals/R-admin.html, ISBN 3-900051-09-7

R Development Core Team (2012b) R internals. R Foundation for Statistical Computing, Vienna, Austria, URL http://CRAN.R-Project.org/doc/manuals/R-ints.html, ISBN 3-900051-14-3

R Development Core Team (2012c) R language. R Foundation for Statistical Computing, Vienna, Austria, URL http://CRAN.R-Project.org/doc/manuals/R-lang.html, ISBN 3-900051-13-5

R Development Core Team (2012d) Writing R extensions. R Foundation for Statistical Computing, Vienna, Austria, URL http://CRAN.R-Project.org/doc/manuals/R-exts.html, ISBN 3-900051-11-9

Runnalls A (2009) Aspects of CXXR internals. In: Directions in Statistical Computing, University of Copenhagen, Denmark

Sanderson C (2010) Armadillo: An open source C++ algebra library for fast prototyping and computationally intensive experiments. Tech. rep., NICTA, URL http://arma.sf.net

Sklyar O, Murdoch D, Smith M, Eddelbuettel D, François R (2012) inline: Inline C, C++, Fortran function calls from R. URL http://CRAN.R-Project.org/package=inline, R package version 0.3.10

Stroustrup B (1997) The C++ Programming Language, 3rd edn. Addison-Wesley

Temple Lang D (2009a) A modest proposal: an approach to making the internal R system extensible. Computational Statistics 24(2):271–281

Temple Lang D (2009b) Working with meta-data from C/C++ code in R: the RGC-CTranslationUnit package. Computational Statistics 24(2):283–293

Thomas A, Redd A (2012) transmission: Continuous time infectious disease models on individual data. URL http://CRAN.R-Project.org/package=transmission, R package version 0.1

Urbanek S (2003) Rserve: A fast way to provide R functionality to applications. In: Hornik K, Leisch F, Zeileis A (eds) Proceedings of the 3rd International Workshop on Distributed Statistical Computing (DSC 2003), TU Vienna, Austria

Urbanek S (2012) Rserve: Binary R server. URL http://CRAN.R-Project.org/package=Rserve, R package version 0.6–8

Vandevoorde D, Josuttis NM (2003) C++ Templates: The Complete Guide. Addison-Wesley, Boston

Venables WN, Ripley BD (2000) S Programming. Statistics and Computing, Springer-Verlag, New York

Subject Index

A
Analysis, 3, 144
Application Programming Interface (API), 19, 22, 155
Armadillo, 15, 54, 75, 139–153, 158, 177, 183, 191, 192, 204
  colvec, 30, 141, 142
  eye(), 151
  mat, 30, 140–142
  randn(), 140
  solve(), 141, 142
  trans(), 140, 151
  zeros(), 151
autoconf, 165

B
B-spline, 169–175
Basic Linear Algebra Subroutine (BLAS), 146, 153, 191
Bell Labs, xi, 3
Benchmark, 10, 17, 20, 115, 124, 143, 152, 180, 183, 191
BioConductor, 59, 192
Boost, 80, 81, 86, 205
  Date_Time, 80, 81
  Python, 83, 86
Bootstrap, 4
Byte-compiler, *see* R, Byte-compiler

C
C, 25, 198
C++
  class, 117–118
  Curiously Recurring Template Pattern (CRTP), 117, 118
  Expression templates, 104
  Functor, 110
  Lazy evaluation, 104, 107, 108
  Operator overloading, 105
  std::unary_function, 110
  template, 110, 117, 118
  Template meta-programming, 104
C++, 195–205
  C++11, 11, 21, 22
  class, 12, 27, 29, 77, 84, 85, 90–98, 150, 201
  Conditional execution, 199
  deque, 203
  Exceptions, 32–35, 142
  Federation of languages, 6, 198
  Function, 199
  Iterator, 158, 202
  list, 203
  Loops, 198
  map, 202
  multimap, 202
  Object-orientation, 6, 150, 201, 200–201
  Pointers, 199–200
  Scope, 197
  set, 202
  Static typing, 197–198
  std::accumulate(), 43, 160, 203
  std::begin(), 202
  std::count(), 203
  std::end(), 202
  std::find(), 203
  std::for_each(), 203
  std::inner_product(), 141, 142, 203
  std::list, 132
  std::map, 58, 132
  std::multiplies(), 43
  std::pop_back(), 202
  std::push_back(), 202
  std::size(), 202
  std::string, 58, 87, 132, 133

std::transform(), 48, 156, 203
std::unary_function, 29
std::vector, 45, 49, 76, 97–98, 132
str::string, 49
struct, 200
template, 29, 75, 78–81, 203–204
Template meta-programming, 76, 78, 178, 203
Cascading Style Sheets (CSS), 137
cfunction, *see* R, cfunction
clang++, 21, 22
Compiler, 19, 20, 23, 26, 130, 195–197
configure, 166
configure.in, 166
Convolution, 26
CRAN, 19, 25, 59, 74, 153, 155, 192
cxxfunction, *see* R, cxxfunction

D

Density estimation, 3, 136
Domain-specific language (DSL), 3
dyn.load(), 25
Dynamic linking, 19

E

Econometrics, 15
Eigen, 75, 177–192
 Array33d, 180
 Array3d, 180
 ArrayXd, 180, 187
 ArrayXXd, 180
 Cholesky decomposition with pivoting, 183
 Column-pivoting Householder QR, 184
 Column-pivoting QR Decomposition, 188
 Full Pivoting LU Decomposition, 186
 GESDD, 191
 HouseholderQR, 189
 Jacobi SVD, 185
 LDLT, 183, 191, 192
 LLT, 189, 191, 192
 Map, 181, 184–187, 189, 191
 Matrix3d, 178
 MatrixXd, 179, 183–187, 189, 191
 MatrixXi, 179
 PivQR, 191
 QR, 191
 SelfAdjointEigenSolver, 185
 solve(), 185, 186
 SparseMatrix, 181
 SVD, 191
 SymmEig, 191, 192
 Vector3d, 178
 VectorXd, 179, 183–187, 189, 191
 VectorXi, 179
Environment variable, 24

F

Fibonacci
 Sequence, 7–15, 25, 31
 Spiral, 7, 8
Fortran, 25

G

g++, *see* GNU Project C and C++ Compiler
gcc, *see* GNU Project C and C++ Compiler
GNU Project C and C++ Compiler, 20, 21, 23, 24, 130, 196, 197
GNU Scientific Library (GSL), 30–32, 75, 155–175, 191, 192
 gsl-config, 30, 166
 gsl_blas_dnrm2, 167, 169
 gsl_bspline, 171, 174
 gsl_const_mksa, 32
 gsl_matrix, 163, 164, 171, 174
 gsl_multifit_linear, 156
 gsl_multifit_wlinear, 171, 174
 gsl_ran_gaussian, 171, 174
 gsl_rng, 30, 173, 174
 gsl_stats_wtss, 171, 174
 gsl_vector, 158–162, 171, 174
 gsl_vector_view, 156
GSL, *see* GNU Scientific Library

I

inline, 9–11, 25–31, 168
 cfunction, *see* R, cfunction
 cxxfunction, *see* R, cxxfunction
Interface, 9

K

Kalman filter, 146–152, 182–183
Kernel estimator, 3, 136

L

Linear Model, 29, 140–146, 156–158, 169, 186–192
Linker, 19, 26, 130, 195–197
LU Decomposition, 186

M

Makefile, 24, 129, 130
Makefile.win, 129
Matlab, 139, 146, 147
Memoization, 12–13
Message Passing Interface (MPI), 134
Modeling, 5

O

Object orientation, 3
Object-orientation, 6, 58–60, 150
Octave, 146
Old Faithful, 3
OLS, *see* Linear Model
OS X, *see* Platform, OS X

P

Platform
   HP UX, 196
   IBM, 196
   Linux, 22, 25, 26, 196
   OS X, 19, 21, 22, 24–26, 196
   Other, 22
   Solaris, 20, 22, 153, 196
   Windows, 19–21, 25, 26, 70, 129, 131, 153, 196
Portable Network Graphics (PNG), 133
Python, 100

Q

QR Decomposition, 184, 186, 188, 189, 191
Quantile estimation, 4, 5

R

R
   .C(), 23
   .Call, 160, 161
   .Call(), 9, 23, 25, 67, 83, 84, 143, 160, 168
   Application Programming Interface (API), 22, 83, 128
   apply, 165
   apply(), 4
   benchmark(), 17
   Byte-compiler, 10, 17, 195
   cfunction(), 25
   CMD COMPILE, 31, 142
   CMD LINK, 130, 197
   CMD SHLIB, 24, 26, 31, 142
   cxxfunction(), 9, 10, 13, 14, 17, 25–27, 29–31, 33, 34, 42, 43, 45–49, 52, 53, 55–57, 60, 76, 142, 143, 151, 169, 185, 186
   density(), 4
   DESCRIPTION, 26, 65, 67, 69, 99
   dyn.load(), 25
   Embedding, 128
   environment(), 99
   generic function, 58
   History, 3
   Inf, 49
   Makevars, 26, 67, 69, 166
   Makevars.win, 67, 69
   NA, 49
   NAMESPACE, 67, 71, 98, 99
   NaN, 49
   Object orientation, 3
   Object-orientation, 6
   package.skeleton(), 66
   pnorm(), 60
   polygon(), 4
   PROTECT, 41
   qnorm(), 196
   quantile(), 4
   Reference Classes, 3, 59–90
   replicate(), 4
   Rmath, 60, 61, 114, 115, 197
   S3, 3
   S4, 3, 41, 58, 59, 86
   sample(), 4, 56
   sapply(), 116
   Scope, 3
   set.seed(), 57
   SEXP, 9, 23, 39, 40, 43, 46, 53, 60, 67, 73, 75, 76, 79, 81, 84, 85, 101, 104, 132, 181, 204
   UNPROTECT, 41
Rank deficiency, 144, 146
Rcpp
   Application Programming Interface, 39–49, 51–60
   as(), 9, 10, 13, 14, 17, 29, 31, 34, 45, 53, 73, 75, 76, 78–81, 84, 85, 88, 95, 133, 177, 204
   attributes, 11–12, 31–32, 172–175
   CharacterVector, 41, 49, 68
   clone(), 46, 48
   cppFunction(), 12, 31, 32
   create, 29
   CxxFlags(), 24, 26
   DataFrame, 29, 55, 60, 173
   Date, 81, 82
   depends, 172
   Environment, 57
   export, 11, 123, 173, 174
   ExpressionVector, 41
   Function, 56, 57, 118
   GenericVector, 41, 52
   IntegerMatrix, 41, 101
   IntegerVector, 41–45, 55, 101, 107, 118
   isNULL, 41
   isObject, 41
   isS4, 41
   LdFlags(), 24, 26
   List, 30, 52–54, 68, 76, 89, 90, 102, 141, 142, 156, 174, 191
   LogicalVector, 41, 48, 106

Modules, 81, 83–102, 135
  class_, 91, 93, 95, 98
  const_method, 98
  constructor, 91, 93, 95, 98
  field, 91, 93
  function, 88, 90
  method, 91, 95, 98
  property, 93, 95
  RCPP_MODULE, 86–88
Named, 29, 51–54, 56, 57, 60, 141, 142, 173, 174
NumericMatrix, 30, 41, 48, 53, 101, 141, 142
NumericVector, 26, 30, 41, 45–48, 52, 53, 60, 68, 85, 101, 104, 105, 107, 115, 123, 141, 142, 156, 169, 191
NumericVector(), 104
operator SEXP(), 77
package, 65–74
Plugin, 30
RawVector, 41, 49
Rcout, 153
Rcpp.package.skeleton(), 65–72, 74, 99, 101
Rcpp.plugin.maker(), 31
RcppArmadillo, 139–153
RNGScope, 56, 85, 115, 123
RObject, 39–41, 191
S4, 58, 96
sourceCpp(), 11, 31, 123, 124, 175
sugar, 103–124
  abs(), 114
  acos(), 114
  all(), 107, 108, 115
  any(), 107, 108, 115
  asin(), 114
  atan(), 114
  beta(), 114
  ceil(), 114
  choose(), 114
  clamp(), 112
  cos(), 114
  cosh(), 114
  cumsum(), 114
  diff(), 111
  digamma(), 114
  Distributions, 115
  dnorm(), 114
  duplicated(), 113
  exp(), 114
  expm1(), 114
  factorial(), 114
  floor(), 114
  gamma(), 114
  ifelse(), 104, 109, 115
  intersect(), 112
  is_false(), 107
  is_na(), 107, 108, 118
  is_true(), 107
  lapply(), 108, 110
  lbeta(), 114
  lchoose(), 114
  lfactorial(), 114
  lgamma(), 114
  log(), 114
  log10(), 114
  log1p(), 114
  mapply(), 110
  max(), 113, 114
  mean(), 113, 114
  min(), 113, 114
  pentagamma(), 114
  pmax(), 109, 112
  pmin(), 109, 112
  pnorm(), 60, 114
  pow(), 114
  psigamma(), 114
  qnorm(), 114
  range(), 114
  rnorm(), 114
  round(), 114
  runif(), 123
  sapply, 118–122
  sapply(), 110, 116
  sd(), 113, 114
  seq_along(), 108
  seq_len(), 110
  setdiff(), 111
  setequal(), 114
  sign(), 111
  signif(), 114
  sin(), 114
  sinh(), 114
  sort_unique(), 112
  sqrt(), 114, 123
  sum(), 114, 123
  table(), 113
  tan(), 114
  tanh(), 114
  tetragamma(), 114
  trigamma(), 114
  trunc(), 114
  union_(), 111
  unique(), 112
  var(), 113, 114
  which_max(), 114
  which_min(), 114
Timer, 179, 180
try(), 32

tryCatch(), 32
wrap(), 9, 10, 13, 14, 17, 34, 42, 43, 45, 57, 73, 75–78, 80, 81, 84, 88, 95, 133, 177, 204
XPtr, 85
Rcpp::Rcout, 179
RcppGSL
CFLags, 166
LdFLags, 166
matrix, 156, 164, 169
vector, 156, 159–162, 173, 174
Recursion, 7, 8, 12
Reference Classes, *see* R, Reference Classes
Reproducible research, 3
Resampling, 4
RInside, 127–137
parseEval, 132, 133
parseEvalQ, 129, 132–134
Rscript, 24, 70, 127, 128, 166
Rserve, 128

S
S, 3
S3, *see* R, S3
S4, *see* R, S4;Rcpp, S4

SEXP, *see* R, SEXP
Shared library, 9
Simulation, 4, 15, 127
Singular-Value Decomposition (SVD), 185, 186, 192
Standard Template Library (STL), 40, 43–45, 54, 76, 96, 104, 132, 141, 158, 198, 201
STL, *see* Standard Template Library (STL)

T
Text file, 127

U
Unit tests, 20

V
Vector Autogression (VAR), 15–17

W
Web application, 136
Windows, *see* Platform, Windows
Wt, 136

X
XML, 137

Software Index

A
Armadillo, 15, 54, 75, 139–141, 144, 146, 151–153, 177, 182, 183, 192
autoconf, 165

B
BLAS, 146, 153, 191
Boost, 75, 80–82, 205
Boost.Python, 83, 86

C
C, ix, 6, 8, 9, 20, 23, 25, 40, 60, 61, 65, 128, 134, 146, 155, 156, 158, 163, 168, 181, 195–200, 204
C++, vii–ix, xi, 4, 6, 8–11, 13–23, 25, 26, 29–35, 40–43, 45, 46, 51–56, 59, 60, 73, 76, 80, 83, 84, 88, 89, 91, 103–105, 109, 114–116, 123, 124, 127, 128, 132–135, 139, 140, 144, 146, 148, 150–153, 159, 172, 177, 182, 187, 189, 191, 195–205
C++, xi, 9, 15, 23, 26, 44, 53, 58, 59, 61, 65, 67–69, 74, 75, 77, 78, 83, 84, 86, 90, 92, 96, 132, 146, 150, 151, 155, 156, 158, 161–163, 166–168, 177, 180, 181, 183, 192, 195, 199, 200
C++11, 21, 31, 42
C#, 40, 203
cda, 102
ceres, 177
compiler, 17
Cpp, 200
CXXR, viii

D
Date_Time, 75, 80, 81

E
Eigen, 75, 177–192

F
Fortran, 9, 25, 168

G
GSL, 155, 156, 158, 160–166, 168, 169, 171–175
gsl, 155, 156
GUTS, 102

H
highlight, 20, 102

I
inline, ix, 9–12, 16, 18, 20, 25–34, 42, 52, 65, 142, 144, 155, 168, 169, 204

J
Java, 40, 58, 59, 203

L
Lapack, 153
Linpack, 144, 146

M
Matlab, 139, 146–148, 151
maxent, 102
mypackage, 67

O
Octave, 146
OpenBLAS, 191

P
parser, 102
planar, 102
Python, 83, 100
Python modules, 86

Q
Qt, 135, 137

R
R, vii–ix, xi, xii, 3, 4, 6–12, 14–26, 29–34, 39–49, 51–61, 65–69, 71–73, 75–78, 81, 83–93, 96, 98–101, 103–105, 107, 108, 111, 112, 114–116, 122–124, 127–136, 139, 140, 142–144, 146–153, 155, 156, 158, 160, 161, 164–169, 172–175, 178–181, 183–185, 190–192, 195–199, 209
RAbstraction, viii
rbenchmark, 10, 20, 142
Rcpp, vii–ix, xi, xii, 3, 6, 9–11, 15, 18–27, 29–31, 33–35, 39–42, 45, 49, 51, 52, 55–61, 65–69, 73–79, 83, 84, 87, 96, 98–100, 103, 104, 114, 115, 118, 127–130, 132–134, 136, 139–141, 152, 153, 155, 156, 158, 160, 168, 172, 177, 179, 181, 185, 192, 204
RcppArmadillo, xi, 15–18, 29, 30, 54, 74, 75, 82, 139–144, 146, 152, 153, 156, 158, 186, 191
RcppBDT, 74, 75, 80–82, 99, 100, 102
rcppbind, viii
RcppClassic, viii
RcppCNPy, 83, 99–102
RcppDE, 53, 54
RcppEigen, xi, 29, 74, 75, 82, 177, 182, 186–189, 191, 192
RcppGSL, 29, 30, 32, 74, 75, 82, 155–159, 161–169, 172, 186, 191
RcppTemplate, viii
rdyncall, ix
RInside, 127–136
Rmath, 196, 197
RObjects, viii
RProtoBuf, xi
RQuantLib, viii, xi
Rserve, viii, 128
RSofia, 102
Rtools, 21
RUnit, 20

S
S, 3, 58

T
transmission, 102

W
wordcloud, 73, 74
Wt, 136, 137

Author Index

A
Abrahams, David, 86, 104, 178, 205
Adler, Daniel, ix
Albert, Carlo, 102
Alken, Patrick, 30, 155
Allaire, JJ, xi
Armstrong, Whit, viii
Auguie, Baptiste, 102
Austen, Matthew H., 201

B
Bachmeier, Lance, 15
Bates, Douglas, viii, xi, 74, 82, 140, 158, 186, 191, 192
Boost, 205
Booth, Michael, 30, 155
Brokken, Frank B., 205
Burns, Patrick, 12

C
Chambers, John M., vii, xi, 3, 23, 58

D
Davies, Jim, 30, 155
DebRoy, Saikat, viii
Diaz, Fernando Cela, 102
Doss, Nathan, 134

E
Eddelbuettel, Dirk, viii, 9, 25, 53, 74, 80, 82, 100, 102, 128, 140, 158, 168, 186, 191

F
Fellows, Ian, 73
François, Romain, viii, 9, 25, 74, 80, 82, 102, 128, 140, 158, 168, 186, 191

G
Gaile, Daniel P., viii
Galassi, Mark, 30, 155
Gentleman, Robert, 23
Google, 177
Gough, Brian, 30, 155
Gropp, William, 134
Grosse-Kunstleve, Ralf W., 86
Guennebaud, Gaël, 177, 183, 192, 204
Gurtovoy, Aleksey, 104, 178, 205

H
Hankin, Robin K. S., 156
Hastie, Trevor J., 58
Hornik, Kurt, xi
Hua, Jianping, 134

J
Jacob, Benoît, 177, 183, 192, 204
Java, James J., viii
Josuttis, Nicolai M., 104, 205
Jungman, Gerard, 30, 155
Jurka, Timothy P., 102

K
King, Michael, 102
Kusnierczyk, Wacek, 10, 142

L
Lajoie, Josée, 204
Leisch, Friedrich, 66
Liang, Gang, viii
Ligges, Uwe, xi
Lippman, Stanley B., 204
Lusk, Ewing, 134

M
Manly, Kenneth E., viii

Matloff, Norman, 23
Meyers, Scott, viii, 6, 198, 200, 201, 203, 205
Moo, Barbara E., 204
Murdoch, Duncan, 9, 21, 25, 168

N
Nguyen, Khanh, viii

P
Plummer, Martyn, xi

R
R Development Core Team, viii, 19–22, 26, 39, 56, 57, 60, 65, 66, 71, 83, 85, 103, 115, 128, 196
Redd, Andrew, 102
Ripley, Brian D., xi, 21, 23, 58
Rossi, Fabrice, 30, 155
Runnalls, Andrew, viii

S
Samperi, Dominick, viii, xi
Sanderson, Conrad, 15, 139, 158, 192, 204
Simon, André, 20
Skjellum, Anthony, 134

Sklyar, Oleg, 9, 25, 168
Smith, Mike, 9, 25, 168
Snow, Greg, 4
StackOverflow, 7
Stroustrup, Bjarne, 204

T
Temple Lang, Duncan, ix
Theiler, James, 30, 155
Thomas, Alun, 102
Tierney, Luke, xi
Tsuruoka, Yoshimasa, 102

U
Urbanek, Simon, viii, xi, 21, 128

V
Vandevoorde, David, 104, 205
Venables, Willian N., 23, 58
Vogel, Sören, 102

W
WikiBooks, 14
Wikipedia, 7, 8, 42, 117, 205

Made in the USA
San Bernardino, CA
22 November 2013